A Requiem
for the Earth
Selected Stories

PAKISTAN WRITERS SERIES

SERIES EDITOR: MUHAMMAD UMAR MEMON

A Requiem for the Earth
Selected Stories

Hasan Manzar

Karachi
Oxford University Press
Oxford   New York   Delhi
1998

Oxford University Press, Walton Street, Oxford OX2 6DP

Oxford   New York

Athens   Auckland   Bangkok   Bombay
Calcutta   Cape Town   Dar es Salaam   Delhi
Florence   Hong Kong   Istanbul   Karachi
Kuala Lumpur   Madras   Madrid   Melbourne
Mexico City   Nairobi   Paris   Singapore
Taipei   Tokyo   Toronto

and associated companies in

Berlin   Ibadan

Oxford is a trade mark of Oxford University Press

© Oxford University Press, 1998

ISBN 0 19 577899 5

Printed in Pakistan at
Asian Packages (Pvt) Ltd., Karachi.
Published by
Ameena Saiyid, Oxford University Press
5-Bangalore Town, Sharae Faisal
P.O. Box 13033, Karachi-75350, Pakistan.

Contents

Editor's Introduction

I

In an interview aired by the Voice of America in 1995, Hasan Manzar remarked:

> My stories are inspired by individuals—ordinary men and women affected by some sorrow, happiness, or longing. When I encounter them, they, as well as the environment in which their reality is embedded, become part of my thinking. Such individuals are not imagined beings. I guess what I'm trying to say is that I'm as far away as anyone can get from any kind of romanticism. Subjects, unless they are firmly grounded in objective reality, leave me cold, and I almost never feel motivated to probe them in my fiction. What moves me, instead, are real, flesh-and-blood people, victims of oppression and violence, scarred by pain and injustice. And it does not matter where they come from.

Clearly, Hasan Manzar's mimetic strategy is pointed away from abstraction, allegorical meaning, or any kind of creative equivocation; rather, it is oriented toward realism in a strict sense. Even deep and knotty psychological problems have to be made evident on the surface, not to be guessed at in the subtle modulations of the psyche well below the surface. Meaning must emerge wearing its own palpable form—gritty, textured, alive to the touch, exposed to the eye. Like Munshi Premchand (1880–1936), whom he resembles closely in his creative vision, if not always in his creative execution, he prefers to cast his stories in the realistic mode, almost never deviating from the traditional geometry of plot and structure. One may find his view of reality a bit too restrictive, for characters in fiction are, to state the obvious, *fictional*. Where they seem real, their reality itself is composite, reconstituted into fictional identity from

fragments of real personalities as seen from the mediating eye
of the beholder. Nonetheless it is a view to which Hasan Manzar
has remained loyal throughout. The dozen or so short stories
which make up the present volume—as well as others in his
three published collections and still others scattered in literary
magazines—never waver in their author's commitment to the
individual, privileging his external reality over probes initiated
into the subterranean and obscure realms of his consciousness.
This individual is so real he could be one's next-door neighbour.
He could be savvy or naïve, rural or urban, educated or illiterate.
He comes from all walks of contemporary Pakistani life, from
every social stratum of society and, as often, from across the
seven seas: Africa, Malaysia, Sri Lanka, Iran, everywhere and
anywhere.

The searing conflicts generated by regional and religious
chauvinism in a highly exploitative society, or even by one's
own unarticulated dark psychological promptings, is the source
from which Hasan Manzar draws the subject matter of his
stories. Among contemporary Pakistani writers, he stands out
for experimentation with the widest possible range of subjects
and for diversity of locale. Pakistan is a multi-lingual and multi-
ethnic society. In no other Urdu writer of this country do these
realities find their most informed and variegated expression as
they do in Hasan Manzar. He touches on most aspects of its
corporate existence, from religious bigotry to corruption at all
levels of society, from regionalism to nationalism. While the
issues are not exclusive to Pakistan, they do have their distinct
Pakistani texture; Hasan Manzar captures it well.

Pakistan was founded on the basis of religious identity and
religion has been present as a major force throughout its fifty
years' of existence, as the principal term of all private and
public discourse. We may well begin with the four stories
'Emancipation,' 'The Beggar Boy,' 'Kanha Devi and her
Family,' and 'The Night of Torment' which offer shades of
spirituality and its misapprehension which often leads to bigotry
and denominational xenophobia—to an exclusivism that feeds
on difference and drives wedges between human beings. But,

again, these stories deal with these issues at the level of the individual. Nowhere does the author single out entire communities for indictment, except in 'The Night of Torment.' In 'Emancipation' a young married Hindu woman, bearing the stigma of barrenness, is assaulted by a Muslim ticket checker in a deserted railroad car one dreary evening during a trip to the Holy Ganges. She manages to walk away from the terrifying ordeal unblemished, thanks to her presence of mind and wit. But for her husband and her in-laws, she is henceforward a condemned woman, a veritable cancer growing on their purest religious honour, a walking reminder of their largely imagined humiliation, an affront to their piety. On the other hand, her assailant's family stands solidly behind him and bend every effort to secure his acquittal, even soliciting help from a 'Muslim holy man.' Worse yet, the assailant's wife, rather than being cross with her husband, raises the accusing finger at his victim.

Hasan Manzar is careful not to exploit the story's negative potential, to underscore generalized views and attitudes about whole groups, or to judge the quiet spirituality inherent in religions by the violent, often explosive, conduct of their adherents. His emphasis remains throughout on how individuals understand and interpret religion, and how they often use it for entirely non-religious ends. The tragic event, and its even greater tragic consequences, turn the victim's thoughts inward, give her the opportunity to reflect on her own attitudes towards the religious other, to distinguish between essential spirituality, with its concomitant potential for inner transformation, and the seductive lure of the exoterica of religion.

This reflection, in turn, becomes a liberating and empowering experience. In the concluding brief segment of the story, set apart from the larger first portion, we meet her not in her native India, but somewhere overseas, with a West Indian friend, as the two playfully hurl coins at empty beer cans floating in a river.

'Another round—shall we?' my West Indian friend asked.

'No, that'll do for today,' I said. 'I've got to cook supper for the children as soon as I get back home. My husband is on call tonight. And since today is my lucky day, I hope the phone won't ring every five minutes for him to rush to the hospital to take care of some drunkard.' (pp. 130–1)

These sentences project a woman confident of herself. Full of verve and brio, she is looking into the future with strength and optimism, in sharp contrast to the meek and trodden housewife of a few years ago who bent over backwards to ingratiate herself with her in-laws, as she wallowed in maudlin self-pity, denying her possibility. She is happily remarried now, has children and—what she didn't have before—an understanding and loving mate, whom she looks forward to meeting with eager anticipation at the end of the day. The narrative is purposely silent about the developments in her life since her and her assailant's acquittal. But none of this would have been possible without a sense of liberation and empowerment blossoming deep within the protagonist's psyche.

Exaggerated confidence in the redeeming and salvafic power of religious exoterica for adherents of one religion or another charged with missionary zeal provides the subject of the short story 'The Beggar Boy.' The contradiction between true spirituality which lies at the core of religion and slavish and hypocritical adherence to the rituals of religion is subtly presented in this story. The entire family of Bare Babu, the station master, is bent on making a Muslim out of Ramsarna, a homeless Hindu boy who comes begging for scraps of food at their house every so soften during the week. With great deftness Hasan Manzar here probes the motives and the conduct of several members of this family to show how their actions are at odds with their avowed scramble for beatitude, whether for themselves or for those outside their fold. These motives are anything but selfless. The oldest boy of the family has no qualms about forcing himself upon a village girl for sex—stuffing a shiny four-anna coin in her hands for services unwillingly rendered—while Bare Babu, the essence of righteousness and virtue, manages to fit into his out-of-town day-trip a quick visit to a prostitute's balcony.

The suggestion to change his faith initially terrifies the beggar boy. But his fear stems more from a feeling of uncertainty and visceral dread of the unknown than from deep conviction about his faith. Like most individuals, he has inherited his faith and grown used to it, without the need to search within himself for its essence or meaning. In fact, it is just this suggestion that forces upon him the need to articulate his religious identity for the first time.

> [C]rossing over to the other bank, he came to a *peepul* tree and prostrated himself in reverential worship before the stone that lay at its foot. For the first time ever, love for his religion had suddenly blossomed in his heart. (p. 153)

He stays away from Bare Babu's household, but only for a day. The pangs of hunger bring him back, and the next day finds him groveling at Bare Babu's doorway again, seduced into the benefits of conversion—'Two meals a day—the same food as we eat, not some rotten bread thrown at beggars' (p. 155), nice clothes to wear, freedom from untouchability, even education. The boy succumbs, as temptation's tidal wave washes away the mud embankments of his hesitation.

Reciting the *kalima* proves easy enough, but the boy panics when the family tries to snip off his *chirki*—the small tuft of hair traditional Hindus wear on a shaven head as a mark of devoutness. Conversion is one thing, losing the *chirki* quite another. He hadn't bargained for it, so he flees. He comes upon a Hindu priest lolling on a raised platform outside the temple in the darkening shadows of the evening. He quizzes the boy and learns about his narrow escape from conversion to Islam. He praises his valiant effort to remain steadfast in his religion. And just as he has seduced him into a promise of better life ('I'm sure some noble soul or the other will take you in' [p. 160]), he makes a sexual assault on him. Ramsarna runs again—twice humiliated, twice betrayed.

In both 'Emancipation' and 'The Beggar Boy' religion is used as an obfuscating cover for one's less than honourable—indeed, entirely un-religious—motives. The incidents themselves are not unique in any way, but in revisiting them in their crass, unredeeming ordinariness, Hasan Manzar seems to be underscoring their great potential for divisiveness and disruption, while subtly alluding to the fact that such conduct is not consistent with true, transformative spirituality which, ultimately, all religions strive for.

The theme of religious difference and its effect on the mind set of minorities is probed with sensitivity and wit in the short story 'Kanha Devi and her Family.' The ideological imperatives of the 1947 partition aside, few can deny that its execution on the human level at least left quite a lot to be desired. Rarely has history witnessed as sloppy and messy a job of uprooting and relocating entire populations. Where Muslims are concerned, scarcely a family was left undivided, some of its members migrating to Pakistan, others opting to live in India. This was less so for the Hindu and Sikh residents of what eventually became West Pakistan. Demographically small, most had managed to emigrate to India. The few who did stay on probably didn't suffer much at the hands of the majority. The main religious conflict in Pakistan appears to have been until quite recently generally inter-denominational, rather than inter-faith. But size and demographic density aside, minority status in itself is hardly a fate to be envied.

Few Urdu writers have focused on the experience of minorities within Pakistan. One looks in vain for a significant story about, for instance, a member of the Parsi community. How does he or she feel in the midst of a Muslim majority? Such exclusion is not only unwarranted but also fundamentally at odds with a fictional logic which assumes, as Hasan Manzar's does assume, the individual to be the principal value and measure of all creative endeavour. One's religious or ethnic identity is only secondary to one's status as a human being.

Whatever importance religious and ethnic questions may assume, they do so obliquely through the lives of individuals.

Kanha Devi and her family are Sindhi Hindus who have stayed on in Pakistan following Partition. This family and a few others live in an enclave surrounded by Muslims. The interaction between the two communities is one of mutual respect and harmony—perhaps a bit too idealistic from today's perspective. The Muslim method of greeting as well as certain deferential customs followed by the local Hindus are unhesitatingly observed by both communities. All the same, this mutual regard does not prevent some of the Hindu dwellers of the enclave and especially Kanha Devi's husband, Chandarmal, and her daughter-in-law, Damayanti, from feeling claustrophobic, jittery, and nervous. Their anxiety grows in proportion to news of atrocities committed against Muslims in India, each fresh assault invoking renewed fear of reprisals from the very people among whom they have so far lived in relative peace. Damayanti, an Indian national married into the family who has never reconciled to her life in Pakistan, loses her sleep, desperately longs to return to her native India, and dangles precariously on the verge of paranoid collapse.

But if Kanha Devi's husband and daughter-in-law live out their days in cloistered marginality, Kanha Devi herself goes about the business of life supremely unaffected by the supposedly hostile presence of the 'other' around her. She is at peace with her environment, fully integrated with it. And so is her son, Kishan Chand, who tells his nagging wife in no uncertain terms:

> You know what I think? If you had been born here, then you too would be like mother: you'd obey your religion and not hate others for obeying theirs. Anyway, why would I want to abandon this country? (p. 48)

And elsewhere:

> Let's just say that I'm in no mood to emigrate to Bharat. I'm happy here. I've grown up among these people and consider them my

own. Your misfortune is that you grew up in an environment full of instigators, people who keep themselves in business by stirring up members of one faith against members of another, and send one caste against the throat of another. Lucky for the politicians! Even in this day and age, they can find enough ignorant people to shore up communal unrest. (pp. 48–9)

Damayanti's unhappiness comes to a head following the latest news of anti-Muslim riots in India and their potentially damaging affects for the Pakistani Hindus. She feels like an animal tied to a stake and beaten to death. She shifts restlessly in bed, unable to sleep.

Kishan, still in bed, threw his arms around a swollen-faced groggy-eyed Damayanti, who was sitting up in bed beside him, and asked, 'What's the matter? Don't tell me somebody drank from your glass again.'

'That happens every day. How much can one avoid ...'

'Then *don't!*' Kishan said, lifting a lock of her hair, and then added, 'Join the others. Mix with them.'

Damayanti freed her neck from his coiling arms and said, 'Come to Bharat with me. I will never be able to sleep peacefully in this country.'

'Why? Do beds have thorns here?'

'This isn't your country. It's *theirs.*'

'Theirs—who?'

'Those who surround us. Who created this country in the name of religion.'

For the next few minutes Kishan strained as though trying to read some invisible writing on Damayanti's face. He said, 'Look at it this way: if a woman can be wife to one and mother to another at the same time, then why can't the same piece of land be held dear by some, because it was gotten in the name of religion, and be respected as a motherland by others? Or must we have two Damayantis?...Only then would it make sense to think of one as mother and the other as wife.' (p. 51)

But it is Parsumal, Kanha Devi's brother-in-law, who more nearly approximates her refreshingly liberal and delightfully ecumenical spirit. He is afflicted with some mental disorder, which comes and goes. A deeply religious man, much given to recitation of the Hindu holy texts, he is nonetheless unsparing in his brutally candid criticism of the hypocrisy in the family. But his tongue-lashing is reserved for those times when he is seized by fits of insanity, when he rants and raves, shouts and screams. Thus 'in madness he manages to tell what is truly blameworthy in man, never that which man has fabricated in order to put some above some others' (p. 44).

In their own ways both Kanha Devi and Persumal represent the finest spirit of religion and exude the reassuring power of religious belief. Genuine religious feeling makes for acceptance, not for hostility, distance and rejection.

In a country where a poor taxi-driver is worried more about the fate of his fare than about arbitrary brown-outs and the disruption of the fresh water supply with a regularity that defies imagination (not to speak of the poor condition of the roads, which directly affects the means of his livelihood), the reiteration of some age-old lessons would appear hardly out of place. 'Emancipation,' 'The Beggar Boy' and 'Kanha Devi and her Family' do just that: dispel some of the obscuring mists from religious exoterica and reestablish the transformative principle inherent in true spirituality, with its emphasis on understanding, open-mindedness and tolerance.

The dark underside of religion, its sable fascination for the masses, and the effects of piety-run-amok on the individual are explored with rare sensitivity and power in the 'The Night of Torment.' The story unfolds in post-revolutionary Iran where the religious establishment feels duty-bound to regulate public morals. A prostitute, mother of four, without protection or support of any kind, and her 'visitor' are whisked away from her cramped two-room apartment in the red-light area by the piety-patrol, the 'cloak-and-rosary men' as they are called, while her children spend the whole of that day and the following night in anxious waiting, trembling now from fear, now from hunger. When she does return the next morning, she can barely walk. She staggers in and collapses. She has been flogged. Deep lash marks are cut into the skin of her back and her shirt clings to her body, a body now etched by the jagged geography of pain. The rest of the story is a numbing recollection of the events of the day before the unforgiving religious tribunal as they impinge on the woman's consciousness, now lucid, now exhausted.

As the scene of her humiliating interrogation unfolds through her faltering recollection, one is struck by the sense of dignity and moral strength about the woman and her uncommon courage in the face of the brutalizing ordeal. All of which make the crass insensitivity of her interrogators—the executors of the divine will—look shockingly appalling and gross.

Whether one should be punished for selling one's body is a complex question and must be asked only after society has ensured all its members of access to what it considers 'honourable' professions, and further ensured that those who lack the required skills will be provided those skills. In other words, the question hinges on volition and necessity. The story's protagonist was not a prostitute for pleasure. She was forced into it by her circumstances. Which makes her punishment (40 lashes) all the more cruel. Even more cruel is the humiliation, derision, and vulgarity she is subjected to during her interrogation. She is asked questions whose answers are already apparent, so that some perverse pleasure may be derived in hearing them from her mouth. But in each instance, she comports herself with uncommon dignity, not even a hint of which can be perceived in her interrogators. Not only are she and her 'visitor' kept hungry for the entire time, her pleas for her children's safety fall on deaf ears. She is asked instead, 'Who is their father?'—rather, to add insult to injury, 'Who are their fathers.' She could have easily identified these men, dragged them to the ordeal as well, especially when none of them was responsible enough to care for the child he had helped bring into this world. 'But I did not want anyone else to go through the agony I was going through' (p. 24).

It is obvious to her that she can't expect even the least bit of consideration from her male interrogators. But what wounds her to her very core is the totally unfeeling attitude of the woman who was appointed to carry out her punishment.

> I looked her in the eye and asked, 'So, it's you who'll...?'
> 'Yes,' she answered.
> I wanted to talk to her, to ask her if she had any children, but by that time my hands and feet had been tied with a rope and secured to the legs of the bench. The woman, devoid of human feelings, a cog in that huge religious machine, was standing over me, on my left side, holding a whip in her hand. (p. 25)

She is finally allowed to leave, but with the exhortation to lead a 'chaste' and 'pious' life. The concluding paragraphs are worth reproducing:

> 'Did you realize the nature of your crime?'
> 'No,' I said.
> Annoyed and angry, the cloak-and-rosary man said: 'The proper punishment for the likes of you is death. You have lost all sense of guilt or shame.'
> I said, 'My lord, a lot more besides the sense of guilt died in me today. But if you really want to know, I never had any sense of guilt.'
> He raised his hand to slap me, but I addressed him with courage—the kind of courage that wells up in those who are at the brink of extinction. I said, 'I feel sorry for you.'
> He held back his raised hand and asked: 'For me? Why?'
> 'For what you are doing,' I told him.
> 'What do you mean?'
> 'I mean that just as men have changed the course of my life, never allowing me to become what I could have, or what any woman could have, in the same way you have brutalized this other woman as well. She should have been rocking a cradle and singing lullabies, but just as you purchased me, you have purchased her as well, and put a whip in her hands.' (p. 25–6)

And all this in the name of religion! The parting admonition to lead a 'chaste' and 'pious' life mocks the very words themselves. She is far more chaste and pious than those who grandly sit in judgment over her. Like any good writer, Hasan Manzar forces us to question and rethink the meanings of words which have been devalued by time, practice, and sheer exploitative usage. 'The Night of Torment' does just that: it throws the two words back at the keepers of public morality as mirrors in which they appear far less than 'chaste' and 'pious.' This story is not an indictment of religion as such. Hasan Manzar is, as he himself states, a deeply religious man. He is not against religion, but rather against its exploitation, against fake spirituality, against the bogus purity that suffocates.

The complex problem of nationalism has not escaped Hasan Manzar's attention, either. But, again, he approaches it from the perspective of the common man. What possibly can words like 'my country' and 'my nation' mean for an underprivileged man who has difficulty putting food on the table for his family? As a concept nationalism has done more to split people along racial, cultural and linguistic lines than to bring them together. Its tracks are still warm with devastation across contemporary human geography. Even in the lands of its invention, it was not achieved without bloodshed. In third-world countries its devastation has been phenomenal, however. Here, it has often meant little more than a diversionary tactic of the politician to disclaim his own responsibility in the plight of the ordinary man. Such a man may be worked to whatever pitch of feverish xenophobia required by politicians, but deep down he knows that his 'country' is where his next meal comes from.

Hasan Manzar takes up this theme in the short story 'A Man's Country'. The time of the story is the period just prior to the breakup of Pakistan (1971) and the emergence of East Pakistan as the sovereign nation of Bangladesh. Most of the action takes place on a ship where the staff is neatly divided into two groups: Bengalis or East Pakistanis, including the protagonist Mobinurrohman—or 'Mobinur,' as his shipmates call him—and the West Pakistanis, among whom belongs the story's narrator. The ship is every bit a microcosm replicating some of the same prejudices and racial feelings of superiority that operated at large among the West Pakistanis vis-à-vis the Bengalis, and the latter's blanket anger with every single West Pakistani for the accumulated humiliations and misfortunes visited upon them by nature and man. The Bengalis are ridiculed and made the butt of cruel jokes on the ship just as in West Pakistani society. Mobinur is thus portrayed as niggardly, stunted, dark, obsessed with money, underfed, aloof, averse even to a little bit of fun (that is, drinks and women), and far too devout for anyone's good. But the writer is careful to give some of these characteristics a credible basis. Mobinur's self-denial and austerity stem from a grinding sense of responsibility to his family: he is the sole provider of a large family back home: a wife, both parents, a sister and three younger brothers. In three years he hasn't once returned home, for the ship's recruiting office is located in Karachi and a trip home would have involved expenses he couldn't very well afford. So the time between signing off a ship and signing back on another he spends in Karachi doing small jobs as a day-labourer to earn some extra cash.

Just prior to the creation of Bangladesh, the situation at home puts Mobinur in a gloomy mood. First the cyclone wreaks incalculable havoc on his family, and then his sister's husband is gunned down by—he tells the narrator—'[y]our people.' So '[n]ow,' he predicts,

> a var [war] must happen. The man on German line seep [ship] tell me all. Thing being really bad. What can a poor man doing? Must fight to save life of vife and sildren [wife and children]. Now nobody can stopping var. (p. 64)

And war does come—in time and irrevocably. But that day the narrator realizes for the first time the searing intensity of Bengali feeling against West Pakistanis.

> That was the day I discovered what the people of Mobinur's race thought of us. It was like what the blacks of South Africa thought of the whites who were few in number but controlled everything in their country—the government, the army, the navy, the air force. In fact, they considered us to be even worse than the whites in South Africa; they had the same view of us as the Uhurus of Kenya had

of the English: they were sent from England to rule them, but after staying there their entire lives and after making money they had ravaged the place before leaving. (p. 66)

When the narrator next meets Mobinur in the shipping office at Karachi, he perceives a palpable change in Mobinur's and his mates' attitude. An air of defiance and insolence surrounds them.

> Passing him by I casually said *bhaalo* to him.
> He said, 'I thinking you not say *bhaalo* to me anymore.'
> 'Why?' I asked.
> 'You vill [will] find out. Let time coming.'
> He and his mates were in a hurry to get to East Pakistan so I couldn't talk to him anymore. (p. 67)

What happens next is history. The war and resulting independence didn't take away the ills from Bangladesh. Driven by dire need and grinding poverty, Bengalis started to enter Pakistan illegally looking for employment. Whenever the narrator ran into one of them, he couldn't resist a feeling of perverse satisfaction at the man's plight, in spite of his deeper realization that where economics were concerned, the fate of the common man in Pakistan was hardly any better. One day he spots the selfsame Mobinur in the shipping office. He rushes toward him and assails him with one question after another. Mobinur answers, the haughtiness of their previous meeting conspicuously absent from his face.

> Then I asked him, 'What brings you to Karachi?'
> 'Coming to signing of a seep,' he finally responded.
> 'But you have your Bangladesh now, don't you?' I sputtered, as crudely and inappropriately as the Hindus and Sikhs in India, I hear, used to say to Muslims, 'You have your Pakistan now. Why don't you go there?'
> Almost vengefully I asked him. 'You didn't go to Bangladesh?'
> He ignored the mercilessness of my tone and said, 'My Bangladesh right here.'
> 'What?' I asked.

He repeated his answer, as if explaining something to an ignoramus:

'My contree—this office, this work, right here.' (p. 69)

Of course no overnight change of heart has occurred in the protagonist to be seduced back into the idea of an undivided Pakistan. If there has been a change at all it is in the sobering realization that 'nationalism' is a hollow concept for the poor citizens of a country.

The colonization of Africa—the so-called Dark Continent—by the Europeans, their insatiable greed for and appropriation of that continent's riches, racial discrimination and such other issues provide the subject of the story 'White Man's World.' The story is narrated by a precocious young white boy who lives with his parents and two siblings in a small African town 'where there is no electricity.' The family is originally from Pretoria in South Africa. Because of its concern for the black population and its warm and humane treatment of them, this family stands out as an anomaly in the atmosphere of injustice and racial hatred rampant among white Africans, including the narrator's own grandmother. An Afrikaner, she never did forgive her Dutch husband who, later in life, took to living with a dark (i.e., Asian) woman and was consequently barred from living in South Africa. She considered him 'low,' not because he lived in sin, but, worse, because he lived with a dark woman. The old lady hates Asians and Africans with a passion, and lives in the cloistered safety of an exclusive white neighbourhood.

The land-grab and general greed of the white man is dramatized for the young narrator in the central question: How much land does a man need? The question etched on the boy's mind after reading Tolstoy's story of roughly the same title lurks subliminally in his consciousness for a long time and breaks up in all its urgency one day as he witnesses, along with his sister Tina, the burial of the Muslim driver of his doctor father's team who has suddenly died of, apparently, natural causes. His father had enjoyed a warm and close relationship with this driver. They had been together on many long trips through the dense forest and shared their food—that's when 'Papa wasn't eating pig's meat' (p. 111). As the driver's dead body is lowered into the grave, Tina whispers into the boy's ear, 'Six feet. See? From his head to his toes, that was all the land he needed' (p. 110).

Just six feet and the insatiable hunger of Pakhom for the Bashkirs' land in Tolstoy's story—emblematic of the Europeans' land-grab across far-flung continents—superimpose on the boy's consciousness as a pair of brutal truths eternally poised against each other. The driver's death, which has touched the boy deeply, puts him in a somber and introspective mood, and a turmoil is brought on in the deeper self as images of the past, focused around the mistreatment of the native black population and the 'dark' Asians, begin to reconstitute themselves with a haunting tenacity. He remembers how, during a train ride, a white woman had taken exception to the presence of a black boy an ailing Asian woman had brought along in the compartment to look after her. How Sunday sermons at the church reiterated the notion of racial segregation with chilling obduracy, as if it were God's own irrevocable will.

He recalls the trip to his grandmother's home the previous summer and what she said to his father at their departure:

> 'Hank, you are welcome to come back here anytime you want. This is the only country in the world where the settlers had only one ideal—that when they sit outside on the steps of their houses, they should not see the smoke rising from the chimney of their nearest neighbour's house' (p. 112).

And later, during the night-long drive from Pretoria to Durban, when everybody had fallen asleep in the car, the hushed conversation between his father and a friend of his, in which they said

> things which were altogether new for me. For instance, they said that when the settlements in the South had just begun, some woman had written in her journal that the policy of equality for blacks and whites was against the teachings of the Bible. [...]

Another thing they talked about was the belief among the white people in the South that it was against God's laws to give equal rights to blacks and whites, and that it was every free man's birthright to acquire as much land as he wanted. [...]

They talked about original people of Australia, about the American Indians, about the Israelis who had come from Europe and about the Jews who were non-European.

I felt as if all the countries of the world had been taken over by the white people, each one of whom was running, like Pakhom, to possess as much land as he could, even if in the effort he had to destroy the Negroes, the American Indians, and many other darker races of the world. (pp. 112–13)

The story ends with the boy, his head buzzing with a complex of troubling thoughts, asking his parents rhetorically,

Maybe, as you say, Tolstoy has written good stories, but I think Tina lies when she says that a man, from his head to his toes, really needs only six feet land. My question is: How much land does a white man need? (pp. 113–14).

Indeed, how much?!

Although somewhat predictable and slightly flawed by an undertone of misplaced joviality and playful sarcasm, 'A Requiem for the Earth' is woven around a very serious subject: the destruction of the earth by man. Phantasmagoric and futuristic, the story underscores the gradual disappearance of woman from the planet and, along with this, that of all beauty and poetry. As a result of man's willful destruction of the world's ecological balance by too many nuclear tests and their radioactive fallout, by the use of insecticides and food preservatives and additives, women in this imagined futuristic time have lost the ability to bear female babies. This has become painfully clear after a thorough scientific study of the female genetic make-up. By the third generation after the onset of the calamity, women have all but disappeared, except for a single woman, the wife of a schoolteacher, in a remote mountain area.

But she is already ill, and is thus moving toward humanity's irrevocable end.

Intensive genetic research has finally succeeded in finding a cure which is expected to reverse the disorder. The global organization called Save Mankind moves into high gear. Several helicopters are dispatched to the region where the last woman on earth is now counting her last days. Plead as much with her husband as they might, he will not let the representatives of the Save Mankind Organization come anywhere near his wife, who meanwhile does die. The husband comes out of his cave and starts digging a grave to bury her.

This chilling prophesy of the end of time and human history as we know it, wrought by man's incurable arrogance, undergirds Hasan Manzar's larger interest in the fate of man at a global level.

Although not narrated on the same global scale as 'A Requiem for the Earth,' the story 'A Tough Journey' is nonetheless just as concerned with humanity's destruction of itself—from within, at the level of ethics and corruption. It is the story of an upper-middle-class civil engineer and his peon, and the unlikely trip the two of them take to the village water reservoir the engineer has been charged with overseeing. As with any large-scale engineering project, the construction of this reservoir involves large amounts of money, government agencies, private contractors, and an administrative bureaucracy that can be exploited by those who control it for their own advantage. And of course, it is the poor villagers whom the pervasive system of graft and kickbacks exploit the most.

For the engineer, such corruption is the natural purview of his middle-class circles, and is seen simply as the way business gets done. For the menial office worker, however, situated as he is outside such middle-class dealings, it is not the system of corruption that is immoral, but rather his exclusion from it. In a troubling yet deliciously ironic twist, the office menial seeks justice not in the elimination of corruption from society, but rather in the fair distribution of graft throughout society, so that it might reach his low station, too, in an 'equitable' way. And the menial is not averse to his own variety of extortion in the service of this larger 'justice.' Piling on the irony, the engineer's anger and frustration at his office worker's extortionist tactics is really just confirmation that the *pattewala* has learned the lessons of corruption well. The tables are turned, and it is the boss who ultimately emerges helpless and exploited. True situational justice? Disgusting, totalizing corruption? To its credit, Hasan Manzar's narrative does not decide for us.

Among stories written along less somber and troubling questions are 'Together,' 'The Poor Dears' and 'The Drizzle.' All three project subtle transformations in their protagonists. 'Together' narrates a kind of emotional dance between an aging widower and his widowed daughter-in-law. It is a story almost absent of any traditionally conceived plot, focusing instead on the tenderness and compassion of the two main characters as they come to terms with each other's emotional wounds. The daughter-in-law has been orphaned at an early age, and has grown up moving from one relative's house to another, never experiencing a sense of *home* in any terms she may have called her own. Even when she marries the old man's son, it merely is a replacement for his first wife, who had earlier left him. And her situation at her in-law's house is made even more difficult when she and her husband are unable to conceive. In a deftly articulated narrative recounting, the reader witnesses the strained affection she shares with her mother-in-law, and although never explicitly mentioned, one cannot help but note the profound, almost existential sorrow that fills her psyche. Yet she is stoic, and comports herself with an endearing and deferential equanimity.

With the passing of his wife—and along with her the luxury of his prior aloofness—the old man finds himself untethered from the routine comfort his earlier domestic life had provided him. He realizes his days are growing short, and is beset with the anxiety of providing for his daughter-in-law after he, too, must eventually leave her. With a supremely light touch, the narrative portrays him as a man called upon to shed the accreted layers of his earlier disinvolvement and to begin in his old age to come to terms with new responsibilites and hitherto unacknowledged emotional priorities.

In a series of feints and misstarts, the two protagonists circle about each other's most deep-seated and painful concerns, producing in each other a new domain of anguish before achieving the ease and equilibrium they so poignantly seek. The subject matter here is delicate and elusive—and, to the best of my knowledge, entirely unprecedented in modern Urdu prose fiction—yet Hasan Manzar's narrative never intrudes or overstates. The reader is shown the transformations of the two characters as much by what is not said as by what is.

In 'The Poor Dears,' the situation of an economically depressed contemporary Pakistani family is seen from the perspective of an expatriate South Asian writer living in London. In all their banality and observance of middle-class appearances, the widowed lady and her two daughters inspire in the reader only the gentlest feelings of regard and understanding. And one comes away feeling strangely nourished by a secret knowledge: wisdom is compassion, compassion being nothing other than the ability to stretch the limits of one's emotional horizon to make room for the other, as the other is, with all his human failings, his particular existential situation. The redeeming moment in the story comes in the transformation of the writer himself. He decides not to send the tape to the widowed lady, yet, paradoxically, nothing would have been more welcome to her family than this tape. The writer has matured enough to realize that to put this family in touch with its past may be an act of charity, but to preserve the family its human dignity may eventually be an act of greater kindness. The real protagonist of the story is thus not the family or any of the many characters the itinerant writer meets throughout his travels in South and Southeast Asia but the writer himself. Against the static and possibly unalterable condition of the family, it is the writer's reality that shows a dynamic change, the melange of individuals and situations encountered as so many signposts along a meandering voyage of self-discovery.

'The Drizzle' revisits the motif of inner transformation with great subtlety and power. Its narrative texture has the feel of austere elegance, at once eloquent and shy. A moment of discovery in which pretensions drop away from the self and one feels magically light and unencumbered, but nonetheless quite privileged and maybe even thankful.

Miss Kamariya, a devout young Malay Muslim woman, from the village of Kedah, works as a maid in the household of a certain 'Madame' who lives in the city. Bulan Puasa—i.e., Eid al-Fitr—is only days away and Miss Kamariya, who is fasting, has planned to celebrate it with her family in her village. Her earlier enthusiasm about the prospect of returning home has considerably dampened as she discovers to her disappointment that the gold necklace, which she had bought after much thrift and economy to show herself off a bit to her village friends, is missing. Although she has no proof, she nonetheless suspects the teen-age Minachi, who comes a couple of times a week to do the laundry at the Madame's. Suspicion comes easily enough: 'Minachi was a Ceylonese Tamil and, what is worse, a non-Muslim, and a Hindu to boot' (p. 5).

To have her produce the necklace, Kamariya thinks up an eloquent ruse: she tells Minachi she has asked the *bomoh*—the Malay medicine man—back in her village to use a spell to flush out the thief, and to make doubly sure has also asked her father to have *Bachaan Yasin* done to find the thief. But Minachi plays the innocent, which only convinces Kamariya that she is not only a thief but, indeed, a habitual thief.

Minachi had stolen the necklace all right. Eventually, though, she restores it to its rightful owner. She says she had a dream in which she was shown where the necklace lay, and leads Kamariya to it. An elaborate ruse, but nonetheless one which results in some minimal measure of face-saving for the repentant. The girl has clearly lost the trust of the family. She has plummeted so far down in their eyes that her presence keeps them on their guards, watchful and suspecting. Madame is even thinking of firing her.

By contrast, Kamariya's own reaction is pleasantly unexpected. What if Minachi did steal the necklace? The important thing was that the light of conscience hadn't entirely died out in her. She could feel another person's unhappiness and sense of loss, and was capable of reflection and transformation.

But eventually it is rather Kamariya's transformation from a suspecting individual into one trusting and full of compassion that the writer—I'm strongly inclined to believe—wants to underscore. The palpable proof of this comes when Kamariya returns from her vacation, bringing gifts for her employer's family. After she doles them out, she asks:

> 'Minachi come Madame?'
> 'Yes,' Madame replied, shaking her head. 'But I am thinking of firing her. Now that you are back, do look for another washing girl.'
> 'Why Madame?' Kamariya asked, astonished. 'I have brought a present for her, too.'
> 'Because she is a thief. She is not a good girl.'
> 'No, Madame, Minachi no thief. She is a good girl.' Kamariya's face was perfectly calm. (p. 10)

Madame is nonplused. Kamariya is a strange girl, she concludes. Kamariya thinks quietly for a while and then says in a halting voice, 'She did steal, but how can she become a bad girl by just one such act? She still has fear in her heart' (ibid.).

As the above discussion of a few of the stories offered here in translation for the reader shows, the experience of disharmony which individuals feel with their environment is more down-to-earth and varied in Hasan Manzar. It stems, as Muhammad Salim-ur-Rahman has pointed out, when his characters, mostly ordinary men and women, find themselves 'in conflict either with the norms of their society, which is to a large extent exploitative, or with the secret and confused promptings of their own psyche.'[1] Conflict, however, doesn't always and necessarily breed destruction. Manzar's fundamental humanity shines through in the bleakest moments; his healthy skepticism, in the end, becomes a source of self-discovery, of gentle, confident wisdom.

The fact that Hasan Manzar much resembles Premchand in his concern for the disenfranchised and disinherited is indisputable. Nonetheless his concern is more wide-ranging than Premchand's and is not restricted by geographical boundaries. Moreover, he has the patience and certainly the right disposition to relax his intense concentration on the subject of a story awhile to notice, and even to enjoy, as well as invite his reader to enjoy with him, some of the minutiae that inevitably accompany human situations. In both writers, however, the preoccupation with victimized people and their fate has often resulted in inadequate attention to the technical requirements of their art, to honing and refining the narrative means to reflect more effectively their larger societal concerns. This is more true of Premchand than of Hasan Manzar. It is hard to avoid the impression that both assign a utilitarian value to the act of writing: literature as a means of social uplift and redress, only in the case of Hasan Manzar this value is supplemented with a strong emphasis on the potential of fictional writing as a source of knowledge and understanding.

Where this concern takes over Hasan Manzar's consciousness entirely, the resulting story carries a hint of a flaw in the medium, a sense of an inadequacy which could have been removed with just a little closer technical attention. This happens invariably in the concluding part of a story and is triggered by insufficient clarity about the role of the character, the narrator, and the writer. The three personae are not, nor should be, identical, though in the hands of a more critically-minded writer an occasional and brief blurring of contiguous boundaries could be creatively exploited to enhance the aesthetic potential.

By way of illustration, take the story 'Kanha Devi and her Family.' The story effectively ended with Persu's response to Damayanti, 'You know something? [...] These pigeons roost in that minaret at night.... More than this I shall not say; in fact, I'm not permitted to say' (p. 52). Maybe even the next paragraph could have been retained to avoid a sense of abruptness and to ease the story into a more natural closure: 'Damayanti wanted to ask who had prevented him, but decided to keep quiet as Persumal had already become joyfully absorbed in his worship' (ibid.).

Kanha Devi's summation of Persu's character sounds moralistic and intrusive, even redundant. The essence of her comment has already been more powerfully conveyed by the graphic image of him feeding the pigeons—pigeons that roost in the minarets of the neighboring mosque.

Likewise in 'The Beggar Boy,' the story has effectively ended before the last paragraph, which only forces to a sense of laboured and contrived closure. The philosophical—or dismissive—tone (it is hard to tell) of Ramsarna's concluding comment sounds like an afterthought, where the writer surreptitiously assumes the identity of his protagonist.

The last two paragraphs of 'The Drizzle,' too, seem to weaken the considerable effect of the story. Their lofty sententiousness seems out of place, lacking in conviction and spontaneity.

Again, the controlled development of the narrative in 'White Man's World' and the strategic deployment of the idea 'how much land does a man need?' at crucial points in the story makes the present ending seem unnecessary. By this time the effect has taken hold of the reader so completely that what he wants, above all, is silence and solitude to reflect, not the intrusion of words that add little. It would have made for more poignant effect had the boy been left with his thoughts, the story's central question rising in crescendo in his brain, rather than enunciating verbally what must of necessity remain voiceless in its silent eloquence.

But if in spite of these minor structural problems the overall narrative architecture holds together—and it does hold together admirably well—the secret lies in Hasan Manzar's vast experience, his microscopic eye that is able to spot stories in the most mundane and commonplace. Eventually his deep human compassion turns the ordinary into an act of sublimity and grace.

II

Because of his ranging and intimate knowledge of national life in its diversity of locale and language, Hasan Manzar may be considered a Pakistani writer in the truest sense. But his embrace of national life in its multiplicity of forms is more the result of an uncommonly inquisitive mind, a vibrant personality, and a vastly compassionate heart, not of a conscious choice to single out for consideration one segment of humanity on the basis of one's national and religious ties with it. Whether an impoverished Sindhi fisherman on a lakeshore in Pakistan or a morally depraved Memon living off his relatives in East Africa, Hasan Manzar makes no distinction in their status as human beings, or on the basis of their religious, cultural, or racial identity. He would have found his subject in any part of the world, and dealt with it with equal sympathy and compassion.

Given the politics of the literary establishment, his own sense of independence and integrity, and his joyous aloofness from literary cliques, this major writer has received appallingly meager critical attention. Hopefully, this collection will contribute in a modest way toward a literary rehabilitation he well deserves and without which we are the poorer.

If critical notice of his work has been meager, information about his life is practically non-existent. The absence of biographical material forced upon me the need to ask him directly for information. Unaccustomed to speaking about himself, he nevertheless graciously obliged, jotting down a few pages of notes, which he sent me in July 1995. The following account is based on those notes.

Syed Manzar Hasan, who signs his creative work as Hasan Manzar, was born on 4 March 1934 in Hapur, a town in the district of Meerut (U.P., India). His father's maternal grandfather had participated in the 1857 War of Independence and so angered the British that they put a price on his head. He consequently escaped to the foothills of the Himalayas and spent the remaining years of his life in hiding. Hasan Manzar spent his childhood in Muradabad, a place famous for its copperwork.

There were many factories and the workers, once a week, would hold a *musha'ira* in one of the rooms, leisurely eating some sweetmeats and puffing on their *huqqas*. Hasan Manzar, against his father's wishes, was admitted in a school where pupils were made to sit on coarse rush mats. He despised the school and started skipping classes, roaming around in the factory area and in the town's crowded side-streets observing life, rather than studying books. Which 'only made father taunt my mother, "I told you, didn't I? What else did you expect?" '

Hasan Manzar was later admitted to Hewett Muslim High School which was located on the bank of Ram Ganga. Across from the school was the small and dingy train station of Katghar, and beyond it farm fields and forests. He remarks:

> I've always found rural life more appealing than city life. And at school I immensely enjoyed plowing, winnowing, watering, and growing vegetables. All the same, I always carried this feeling within me that we somehow didn't belong here. Eventually when we did return to Gorakhpur, where the family had settled from the time of my grandfather, I felt myself a perfect stranger there as well.

Pakistan came into being just as Hasan Manzar was training for the Indian Civil Service. During the 1946 elections he would often be sent to *burqa*-clad women to canvass among them for the Muslim League. But until then he had no idea that soon he'd have to leave India. 15 August 1947 rolled along. He remembers its arrival well:

> There, under the shade of *peepal* trees in the school yard, we chanted *'Hindi hain ham vatan hai sara jahan hamara.'*[2] Later, after I'd migrated to Pakistan, I asked my schoolmates, 'what did you chant on 14 August?' I'd guessed right: they had chanted *'Muslim hain ham vatan hai sara jahan hamara.'*[3] What a fine poet he was, God bless him! He served two warring powers equally well. Anyway, we'd become Pakistanis overnight, but somehow it was hard to get rid of the feeling of being strangers in this new country of ours.

Hasan Manzar graduated from high school in Lahore and studied for the next three year's first at Foreman Christian College and then at Islamia College. Subsequently he enrolled in King Edward Medical College and earned a degree in medicine. He moved to Karachi and took a job in a hospital where the majority of his patients were Shidis working in the coastal salt factories. 'As I lay down in my room at night to sleep,' he vividly recalls,

> I'd hear them sing in their deep, throaty voices—voices which seemed to spread out over the sea and the desert. A sense of immense peace and solitude would settle over me. Sometimes at night I'd go and sleep over the low guard-wall of the footbridges on the paths leading to the sea and whenever anyone tried to scare me with robbers, I'd retort, 'They stand to lose if they kill me. Think about it, what other doctor would come and treat them?'

Subsequently he worked as a surgeon on a Dutch merchant ship and as in-charge of a hospital which specialized in epidemic diseases. His next employment was as Junior Obstetrician and Gynecologist in Saudi Arabian hospital where his boss was an Austrian who had worked in the air force during the Second World War. The training he got here benefited him a great deal when he later worked in bush Africa. On touring health centres and yaws and leprosy clinics in Nigerian jungles he'd often wander into mission hospitals to greet the sisters who ran them and some of them would take advantage of his presence and ask him to perform a minor operation, offering a glass of some fruit juice in hospitality. His fascination with the open life of the wilderness continued even in the jungles and bush-country of Africa, of which he has visited many countries. This itinerant existence has given him this wonderful ability

> to consider every place my own, the wishes and longings of its people my wishes and longings, their struggle for independence my own struggle. I've also been to many Middle Eastern and West European countries. There was a time when I didn't feel much interested in Pakistani political news because my sympathies were

sort of universal. But the precarious situation of the country eventually induced me to take interest in the local news too. How can I not? How can I shut my eyes to the murder and bloodshed that's going on around me since 1987 and sit back smugly?

For his higher studies he went to Scotland, earning two separate post-graduate degrees in Psychiatry from Edinburgh. But he didn't return to Pakistan when done; instead he went to Malaysia, fell in love with its red earth, 'as much as I had earlier fallen in love with the earth of Kenya, Tanganyika, and Zanzibar,' and suffered from it equally. He contracted asthma. He taught in the Department of Psychological Medicine at the university in Kuala Lumpur and also carried on his medical research. His deteriorating health eventually forced him to return to Pakistan, where he didn't settle in Karachi, but rather

I once again chose a city which was considerably smaller and was in the close vicinity of jungles and agricultural fields. I'd absolutely no difficulty embracing the Sindhi peasants as my own, and neither did they consider me a stranger. I wanted to build a *sara'e,* a lodge, for my poor patients who came from rural areas for treatment, because they felt diffident or lacked the means to stay in hospitals or hotels. The same week in May 1990 which saw the completion of my twenty-room *sara'e* also made me a *mohajir* all over again.[4] Up until 1947 I was aiming to become an I.C.S. and would have lived in my town and in my province as one who belonged there. In 1990 I was trying to live like a Sindhi and would have lived as any other Sindhi, but the turn of events pulled the rug from under my feet. Be it as it may, the fact is: neither did my patients let go of me, nor I of them to contemplate yet another move to, say, Karachi or Punjab, though I might add here that I'm not a writer whose geographical boundaries are fixed.

About his writing:

Characters come to mind, or situations, never their national or ethnic identity: American, a resident of U.P., Punjabi, Sindhi, Makrani, Arab, Scott, Black African, Gujarati, Bengali, or Pathan—it makes no difference. Actually, such categories only help identify them in

the colour-scheme of life, but none of the categories—whether
Methodist, Catholic, Shi'a, Sunni, Agakhani, Malay, Chinese,
African—helps us recognize the reality of one who inhabits the
perishable clay body. It doesn't answer the question of how this
man looks from the inside. You can situate a story in any country,
in any land. It's possible that had I lived in the west of Uttar
Pradesh, like R.K. Narayan, I too might have churned out story
after story about a certain 'Malgudi.'

I've been writing all along and have had to abandon a goodly
number of stories half-finished because of the demands of my
profession, which often leaves me little or no time. Then too, the
drafts keep getting misplaced or lost all the time. In the first days
I paid no attention to getting my work published; now, though,
I don't mind sending them out. I've piles of notes which I took as I
examined my patients, riding my bike or, later, driving the car. I've
this rule about writing which I stringently follow: What I'm writing
today should be my best. So, in a manner of speaking, if I'm in
competition with anyone, at all, it is myself.

He credits his mother for his thirst for good books:

Most classics my mother had related to me in the form of stories
already when I was little, and she is also responsible for developing
in me a taste for good movies as well as a fondness for religion.
My father did the same, only on a larger scale. My wife is also very
fond of reading good literature.

I'm a religious man. Painfully shy and reclusive. I easily lose
my way in cities, though not in forests and mountains. A field of
my own, some animals, and right beside them my clinic—this is
what I've always wished for. But now as my powers decline the
thought of such an idyll drifts farther and farther away.

If during a psychotherapy session a patient happens to refer to
one of my fictional characters or to a story I've written, I take it as
a sign of my success. But I usually don't react to such allusions, to
preserve the spirit of the session. Once again the discussion returns
to my patient's fear and his or her problem. This keeps either of us
from cheating the other.

Most of my acquaintances, relatives, patients, and fellow-doctors
do not suspect me of having literary aspirations. I strive for the
simplest and most direct expression in my stories and avoid

romantic speech. I also consider show-offiness of any kind or the desire to impress the reader a taboo. I search for four things in my writing: beauty, truth, purpose, and effect.

Whether or not I'm satisfied with my stories is not an important question for me. I rather take comfort in the fact that to this day I have never written anything in an irresponsible manner. I always avoid writing about something which I suspect might cause disruption, excite baser feelings, or promote ignorance, or, finally, which may smell of any kind of hatred or prejudice. I know Urdu, Hindi, English, Persian, Arabic, Sindhi, and Punjabi to varying degrees of competence and can also get by in a few others. The primary function of language in my view is to bring two people closer. I'm surprised when people, especially the intellectuals, use language as a the basis for spreading hatred and unrest.

Hasan Manzar's first collection of short stories was *Riha'i*. It was published in 1981, followed by *Nadeedi* in 1982. The third, *Insan ka Desh* came out in 1990, and a fourth one, *Soo'i Bhook*, is in press. He has also published, in 1990, an Urdu translation of Munshi Premchand's last and incomplete Hindi novel *Mangal Sutra*, with a substantial Foreword. These days another translation of his—Shivrani Devi's Hindi biographical work *Premchand Ghar Men*—is appearing in installments in the monthly *Afkar*. Hasan Manzar holds Premchand and poet-activist Hasrat Mohani (1875–1951) in great esteem for the fine quality of their writing and for their uncommon humanitarian spirit.

But, he admits, he has also learnt a great deal from Ghalib, Premchand, Pushkin, Tolstoy, Chekov, Gorki, Dostoyevski, and Tagore. 'Even now I can spend any amount of time in their company. But this is only an incomplete list. There are many other writers who have given to me equally substantially.'

Talking about sports, he says:

I've always participated in sports but was aware of possible accidents and danger, and so have always managed to keep a respectable distance between, say, a particularly strong strike of a cricket or hockey ball and myself.

He and his wife, who is a pediatrician, have three children, a son and two daughters. The son is studying to be a psychiatrist; the older daughter is a neurologist and the younger one wants to be a psychologist, 'but we shell see.' His son is married and has given him two granddaughters. 'That's about it,' he concludes. 'Not much to tell, is there? Oh, yes—mine was an arranged marriage, sort of.'

'Some of my characters drink, but I don't,' Hasan Manzar remarks.

If I end up in paradise, I wonder what sorts of arrangements will have to be made for me. I'm diabetic and suffer from high blood pressure. There is no way I can survive on the celestial diet of 'honey' and 'milk'. God knows best!

A question is frequently asked of him: How has psychiatry impacted on his writing and how has his profession helped him in his creative work?

Well, the answer is simple: none at all. Hasan Manzar the writer is a whole lot older than Hasan Manzar the psychiatrist. You see, I'd stared writing already in my early youth. I'd read it to members of our household and to neighbourhood women. My incarnation as a psychiatrist, however, didn't come about until 1970. By then I'd already published quite a few stories. When I'm writing, my patients do not try to force themselves into the group of characters I've seen in the wider field of life or who have emerged in my mind as a consequence of some event or simply of a powerful emotional feeling. Psychiatry, in a manner of speaking, works as a protective barrier at the time of writing. As if each of my patients, male or female, is begging me: Please don't write about me! Please don't expose me! When I've never done this to my friends, how could I do this to my patients?

His first short story was called 'Dihqan'. It came out in *Istiqlal*, a weekly published by the West Pakistani Government. The second story he wrote was, he confesses, quite by accident, 'symbolic'—at least it looks that way now. It was entitled 'Do Sarken, Do Kinare' and came out in 1950.

I've also had my fling with the film world. But I crossed that bridge already during my college years. I retreated and came back. My screenplay was liked by two directors who couldn't be more different: one was learned but less known, the other very well known but insufferably vulgar. Anyway, they both liked it very much. But the former couldn't find the money to finance it, because the heroine of my play was a woman in her 40s. The latter was so impressed by my hold on the details and technique that he wanted me to try out my hand at copying some American film. The idea was to start out with it and after about one-fourth into it switch to my own script. Well, I felt sick to my stomach. Henceforward, I started to pay more attention to my medicine books. Then again, I'd never have survived in the film world. It was not my cup of tea.

He has two incomplete novels and a finished but unpublished play. Also a screen play. 'When will I get to them? I don't know.' Additionally, a book of children's stories is ready and awaits publication.

To me the *raison d'être* of art is that it enables one to admit others into his or her experiences, his or her feelings. This a writer or a poet achieves singly, in his isolation, and in theatre and movies, collectively by an aggregate of players and actors. It is precisely this ability to reach out to others that makes Tolstoy such a great writer and Ghalib such a great poet.

On to a different subject: religion, in my eyes, carries within it a tremendous potential for bringing people together. If it is used hypocritically to sew seeds of divisiveness and fissure, the fault lies squarely with the crazy people who, unable to offer it as a gift to others, hurl their own hatred at them instead. In a way, I consider all men part of a single brotherhood.

III

Of the stories included in this volume, 'A Requiem for the Earth' ('Zamin ka Nauha'), 'The Night of Torment' ('Bipta ki

Raat'), 'Kanha Devi and her Family' ('Kanha Devi ka Gharana'), 'White Man's World' ('Safed Aadmi ki Dunya'), 'The Drizzle' ('Boonda-Baandi') and 'Emancipation' ('Riha'i') are taken from the author's first collection *Riha'i* (Hyderabad: Aagahi Publications, 1981), pp. 1–19, 20–40, 41–57, 78–101, 102–19 and 120–45, respectively; 'The Poor Dears' ('Bechare') and 'The Beggar Boy' ('Ramsarna'), from his second collection *Nadeedi* (Hyderabad: Aagahi Publications, 1982), pp. 1–22 and 119–46; and 'Together' ('Saath'), 'The One Upstairs' ('Uparwali'), 'A Tough Journey' ('Ek Bhari Safar'), and 'A Man's Country' ('Insan ka Desh') from his third collection *Insan ka Desh* (Lahore: Qausain, 1991), pp. 32–49, 50–62, 84–102, and 103–24. 'The Cactus' ('Kektas') appeared in *Aaj* (Summer 1990), pp. 25–30. Some of these stories have earlier appeared in the *Annual of Urdu Studies* 10 (1995) and are reproduced here with permission from the *Annual's* editor. All non-English words appearing in the text have been approximately spelled and italicized.

I owe a debt of gratitude to two friends without whose immense and unflinching help I might not have been able to put this volume together. Faruq Hassan of Dawson College, Montreal, most graciously accepted my request to translate some of the stories. Griffith A. Chaussée not only contributed two translations in the midst of a very heavy schedule of engagements and prior commitments, but also read through the entire work and offered many valuable suggestions for improvement of idiom and expression. Muhammad Salim-ur-Rahman deserves a special note of thanks for contributing a translation to this volume at a time when his literary interests find greater satisfaction in creative writing, editing and reading.

Finally, Hasan Manzar, a warm and kind friend, promptly supplied whatever information I asked of him. Many, many thanks to him.

Muhammad Umar Memon
2 June 1997

NOTES

1. 'Always a Sticky Wicket,' in *Pakistan Times* (Lahore), 27 December 1982.
2. A line from a poem by Muhammad Iqbal (d. 1938); it means: 'We are Indians; the entire world is ours.'
3. *Ibid.,* the line means: 'We are Muslims; the entire world is ours.'
4. Hasan Manzar is alluding to his magnificent and spacious house and the adjacent facility for his patients which he had built in Gulistan-e-Sajjad in Hyderabad from which he was forced out at gun-point following the Sindhi-Mohajir clashes in Hyderabad.

Glossary

Introduction

Bachaan (from *bacha:* to read) *Yasin:* reading/recitation of the
  36th chapter of the Qur'an.
bhaalo: a salutation or greeting common among Bengali people;
  something like 'You're all right?'
bomoh: Malay medicine man.
burqa: a two-piece wrap-around used as a veil and cover for the
  body; used by Muslim women in public.
chirki: a small tuft of hair left on shaven head by Hindus as a
  sign of religiousness and piety.
huqqa: a hubble-bubble, water-cooled pipe.
kalima: Muslim profession of faith ('There is no god but Allah;
  Muhammad is the Prophet of God'); this is the major
  kalima, but there are a number of others in which a Muslim
  testifies to his faith and belief in all other Prophets,
  revealed books, angels, and the Day of Judgment.
mohajir: an immigrant; here, a Muslim who migrated to Pakistan
  following India's Partition in 1947.
musha'ira: a poetic symposium or competition in which poets
  declaim their compositions before an audience.
pattewala: a peon; a man who runs errands.
peepul: the Indian tree, *Ficus religiosa*, or Bo tree; considered
  sacred by many Hindus, often associated with temples.
sara'e: a lodge; inn.

The Drizzle

adek: a younger brother or sister (pronounced without the final *k*).
amah: maidservant.

Bachaan (from *bacha:* to read) *Yasin:* reading/recitation of the 36th chapter of the Qur'an.

Bint (Arabic): daughter (of).

bomoh: Malay medicine man.

Bulan: month; *Puasa:* fast, fasting (here, the Muslim month of fasting: Ramazan).

buka (to open; take off) *puasa* (fast): the meal at sundown with which the daylong fast is terminated, for which the usual word in the subcontinent is *iftar*.

chempedak: jack fruit.

Chi: Miss; Mrs.

Eid: (see under *Hari Raya Puasa*).

dhobi (Hindi): a washerman or washerwoman.

Hari Raya Puasa: Id al-Fitr, the Muslim festival which comes at the end of the month of fasting, Ramazan.

iftar: (see under *buka puasa*).

kakak: elder sister (pronounced without the final *k*).

kampong: village.

kathal (Hindi): jackfruit.

mandi: bath; to bathe.

manggis: mangosteen.

rambutan: same as *manggis*.

suhur (Arabic): the meal, taken in the wee hours of the morning, which commences the Muslim fast.

Sura Yasin: the 36th chapter of the Qur'an; it begins with the word 'Yasin'.

susu: milk.

Night of Torment

Ciao (pronounced Chiao): an Italian interjection, colloquially used for: hi!, hello!; good-bye!, so long!

chilo kebab: grilled chunks of skewered meat and vegetables.

Shariat: Islamic religious law.

A Tough Journey

imam: a Muslim well versed in religious knowledge; also a prayer leader; a distinguished and foremost individual.
hadith: the sayings and reports about the conduct of the Prophet Muhammad.
L.D.C: Lower Division Clerk.
mali: a gardener.
paraathaa: a multi-layered flour pancake fried in clarified butter (*ghee*).
pattewala: a peon; a man who runs errands.
put-vaangoon: son.
sahab munjhna: My Dear Sir.
saa'iin munjhna, munjhna saa'iin : same as above.
salaams: Muslim greeting offered by raising the right hand to the forehead and uttering '*as-salamu alaikum*' ('Peace be on you').
sheedi: a term pejoratively applied to denote somebody of short stature and dark colour, with short and very curly hair.
sheesham: the tree *Dalbergia sisu*; its dark reddish-brown hard wood is used for making furniture.
U.D.C.: Upper Division Clerk

Kanha Devi and Her Family

achaar: oil-based spicy single-or mixed-vegetable pickle.
Allah: the Muslim word for God.
azan: Muslim prayer-call given from the minaret of a mosque five times a day just before the commencement of the ritual prayer.
Bhagavad Gita: name of a philosophic poem (an episode of the *Mahabharata*) held in high regard by the Vaishnavas.
Bhagavata: same as above.
bindiya: a dot, a mark; here, ornament for the forehead usually worn by Hindu women.
chacha: a paternal uncle.

dharma: roughly 'religion' or 'socio-cosmic law'; also: ordinance, statute, law, order, rule, usage, practice.

Divali: Hindu festival of lights.

Holi: spring festival of the Hindus, celebrated at the approach of the vernal equinox by sprinkling colored powders and dyes on each other.

insha'allah: 'If God wills'; used principally by Muslims.

Ishwar: Hindu word for God.

paan: betel; a rolled leaf containing a chewing mixture of crushed betel nut, spices, cured lime and catechu pastes, and often tobacco.

papars: spicy paper-thin wafers made with any of several kinds of lentils; served usually fried or broiled.

salaam: Muslim greeting offered by raising the right hand to the forehead and uttering '*as-salamu alaikum*' ('Peace be on you').

sari: a long piece of cloth wrapped round the body and passed over the head; worn by women.

Shudra: member of the fourth and servile Hindu caste.

ta'wiz: an amulet or charm to ward off evil.

wanyas: a caste of Hindus; in literature and common parlance stands for money-lenders, with a propensity for ruthlessness and exploitation in collecting what is due to them.

A Man's Country

bhaalo: a salutation or greeting common among Bengali people; something like 'You're all right?'

hilsa: the *hilsa*-fish, *Clupea alosa*.

Together

Amma: mother; something like 'Mom'.

Abba: father; something like 'Dad'.

Bitya: daughter, used with a feeling of affection.
daal-bhari roti: *roti* (see below) stuffed with any of the varieties
   of cooked lentils and fried in clarified butter.
kurta: long, loose, collarless Indian shirt.
paisa: one-sixty-fourth—and now one-hundredth—of a rupee.
roti: thin, round, baked bread.

The Poor Dears

bhikshu: one whose subsistence comes from alms; a religious
   mendicant.
maths: a hut or monastery or college for Buddhist or Hindu
   ascetics and monks.
Mira *bhajan*: *bhajan* is a type of devotional song which uses
   bhakti notions and imagery. The name of Mira Bai, the
   famous woman saint-singer of North India who died in
   1547, is closely associated with this poetic genre, which
   she used to express her ardent love for, and devotion to,
   Lord Krishna.

White Man's World

neem: a tree belonging to the *Azadivacta Indica* family, with
   bitter, berry-shaped yellow fruit; its twigs are used as toothbrushes and its leaves, in place of moth-balls to store
   woolen clothes.

Emancipation

Bhagwad Gita: name of a philosophic poem (an episode of the
   Mahabharata) held in high regard by the Vaishnavas.
Bhagwan: Hindu term for God.
Baisakh: the first of the Hindu months falling between April
   and May.

ghat: a wharf, quay; steps leading to the water's edge.

Hare Om: *O Om!*—a pious exclamation used by Hindus; *Om* represents the union of gods Shiva, Vishnu and Brahma; the phrase is often used as a pious salutation.

jadu-tona: magic; spells; voodoo.

Mataji: Mother.

rishi munis: a *rishi* is a singer of sacred hymns; bard; saint-sage; anchorite; a *muni* is an ascetic who has attained a more or less divine nature by mortification; together: religious holy men.

shalwar: women's trousers; relatively narrow around the ankles but quite loose and baggy in the middle; worn with a long shirt (*kurta* or *qamis*); also worn by men in Pakistan.

The One Upstairs

burqa: a two-piece wrap-around used as a veil and cover for the body; used by Muslim women in public.

desi: native, as opposed to foreign.

falsa: the fruit of the *Grewia asiatica*; it is purple in colour.

ghee: clarified butter.

mullah: a Muslim individual well versed in religious knowledge and texts.

The Beggar Boy

anna: formerly, a copper coin, the sixteenth part of a rupee.

babu: Mr, sir; an educated person; a clerk.

Bahu Ji: daughter-in-law; the 'ji' is suffixed to personal names and titles for respect.

Baji: older sister.

Baqar Eid: the Festival of Sacrifice which follows the culmination of the Hajj; second major festival of Islam.

Bare babu: Older or Senior Sir; here, the head of the family.

champa: a kind of fragrant yellow flower.

chirki: a small tuft of hair left on an otherwise shaven head by Hindus as a sign of religiousness and piety.

Dhat, Dhat: an interjection (stop! stand! halt! sit!), to encourage an elephant, or to stop it.

Eid: usually the Muslim festival which terminates the month-long dawn-to-dusk fast.

hauloo: a fool; an idiot.

idgah: a large space (may or may not be enclosed) where the prayer on the two major Muslim festivals of Eid takes place.

kalima: Muslim profession of faith ('There is no god but Allah; Muhammad is the Prophet of God'); this is the major *kalima*, but there are a number of others in which a Muslim testifies to his faith and belief in all other Prophets, revealed books, angels, and the Day of Judgment.

laihnga: long, flared skirt, secured around the waist with a cord.

La ilaha il-lal-lah: 'There is no god but Allah'; Muslim profession of faith.

Magh: name of the tenth Hindu month.

Maharaj: supreme sovereign; title indicative of extreme respect and reverence.

milad: celebration of the Prophet Muhammad's birthday; among South Asian Muslims a gathering, mostly of women, at which religious poetry is chanted, offered as an act of piety.

neem: a tree belonging to the *Azadivacta Indica* family, with bitter, berry-shaped yellow fruit; its twigs are used as toothbrushes and its leaves, in place of moth-balls to store woolen clothes.

paisa: one-sixty-fourth—and now one-hundredth—of a rupee.

paan: betel; a rolled leaf containing a chewing mixture of crushed betel nut, spices, cured lime and catechu pastes, and often tobacco.

panwari: a *pan*-seller.

peepul: the Indian tree, *Ficus religiosa*, or Bo tree; considered sacred by many Hindus, often associated with temples.

pur: city, town; usually added to the name of a city or town

Ram: short for Ramchandarji; the hero of a great Sanskrit epic, the *Ramayan*; an incarnation of the great god Vishnu.

A Requiem for the Earth

chhinka: a suspended woven basket for keeping foodstuffs out of the reach of cats, dogs, and other animals.

kebaya: a long outer garment worn by Malay women.

The Drizzle

It was *bulan puasa,* and Miss Kamariya, daughter of Muhammad Yusuf, was observing the fast.

She was ironing Madame's and the children's clothes in quiet absorption. Behind her in the bathroom, at the tiled washbasin, Minachi, the Hindu Tamil girl, was squeezing the washing in Surf suds. She, too, was quiet. The children had already gone to school, Madame, to shop, the master of the house, to work; only the little girl was around, but she was playing outside in the yard. The house was filled with an eerie silence.

When Minachi emerged from the bathroom with the washing, beads of perspiration glistened on her dark, beautiful face, which she tried to wipe against her shoulder. Although they had stopped talking to each other two days ago, Minachi, passing Kamariya, stopped briefly and asked, 'Any news of the locket?'

Kamariya's hand, which had started to move unusually fast at Minachi's approach, suddenly went slack. She, too, raised a shoulder to wipe the sweat off her tawny face.

The air felt confined and stuffy. Outside, the clouds seemed to have crowded into a part of the sky and gotten stuck. Through the window one saw the leaves of banana and *kathal,* without a stir.

Kamariya raised her head and, staring hard into Minachi's eyes, said, 'No, none at all! But I have asked *bomoh* to help in this matter.'

Minachi wanted to lower her eyes; instead, she picked up courage and asked, 'Oh, and what does he say?'

A few soapsuds remained on her elbow, and water was dripping from the washing.

Kamariya thought deeply and said, 'Just that he will use a spell to find the thief, and then he'll punish him.'

Minachi's face turned blank. She went on to hang the washing outside on the clothesline.

But Kamariya continued, just so Minachi would take note: 'I've also written to my father, and he has promised to have *Bachaan Yasin* done at our house in Kedah. By the time *Sura Yasin* is completely recited, the thief's condition will begin to deteriorate.'

Minachi was quite young, about fourteen or fifteen years old, scarcely the age at which to know the art of suppressing excitement. She walked back and stood right beside Kamariya and asked, 'What is this *Sura Yasin*? And how does one go about setting up this *Bachaan*?'

Kamariya explained: 'My father is very learned in religious knowledge. He will recite the 36th chapter of the Koran. The *imam* of our village mosque and thirty-eight other people, altogether forty of them, will recite the 36th chapter. And you, *dhobi*, just you watch, the thief simply won't be able to get away in the face of all these men doing the *Bachaan*.'

Kamariya's last sentence betrayed bitterness and distance. Never before had she addressed this Tamil Hindu girl as *dhobi*, having only used *adek*, 'little sister.'

Minachi wanted to say something but she choked on her words.

Kamariya concluded: '*Bachaan Yasin* never fails to have effect, *dhobi*, no matter how big the theft. And for a piddling theft, like this one—the local *bomoh* is good enough for that. He says he will take care of it himself: use his magic and track down the thief.'

Coming out of the house, Minachi picked up her bicycle, wrapped her sari tightly around her legs, got on the saddle and said as she rode away, '*Chi Kamariya*, I'm done for the day and I am leaving now.'

Minachi had never before addressed Kamariya as *Chi Kamariya*, 'Miss Kamariya,' but only as *kakak*—the respectful way of addressing someone who was like an elder sister.

Kamariya went on ironing quietly.

Bulan Puasa was coming to an end; and Kamariya had already received permission from Madame to take a few days off to celebrate the festival of *Hari Raya Puasa* in her village with her family. She was planning to fast the last two or three days with the others at her home. No matter with how much deference and affection the people here at Madame's treated her, the unbounded joy of *suhur* and *baka puasa* that she experienced in her own village had no parallel here. Here, fasting was merely an act of worship; there, among her own, to wake up for the wee meal before dawn, when it was still quite dark outside, and the assault of the children on the sweets at *baka puasa*, transformed it into something more than worship. Add to that the joy of Eid itself: new clothes and cosmetics which she would bring with her from the city and which would set the hearts of the village girls ablaze with the desire to seek their fortunes in some big city—to leave their small, quiet world surrounded by rice paddies and coconut palms and step into another, limitless world into which Kamariya, Bint Muhammad Yusuf, had gone empty handed and from which she returned every time loaded down with all kinds of fancy things.

But these past few days the thought of returning home came to her with a shock, as if she were taking a step and finding nothing solid underfoot.

Where was the locket? And where was the chain? Both were made of gold; she had bought them after tremendous sacrifices. A few days ago she could not have imagined returning to Kedah without them. But today, it seemed that the gold chain which had earlier adorned her neck had now somehow slipped down, around her feet, and was keeping her from going home.

She had last seen the locket and the chain just four days ago. After taking a bath, she had put the necklace on and examined herself in the mirror. The locket was substantially heavy and beautifully carved, the chain was delicate and supple; the piece gave her neck and bosom a look of resplendence.

After that, as far as she could remember, Kamariya had put it back in the drawer and had carefully hidden it under the scarves, over which she had scattered a few makeup articles, imitation rings and hairpins.

But the very next day, early in the afternoon—when Kamariya had given the little girl her bath, fed her and put her to sleep, while Madame was taking a nap upstairs and the whole house was plunged into silence—she had an impulse to look at the necklace again.

Quietly she removed her hand from under the sleeping girl's head, got up from the bed without making the slightest sound and tiptoed all the way to the table. She even opened her own drawer as quietly as if she were stealing from it. But within a couple of minutes she flung caution to the winds and began madly rummaging in the drawer. She went over every article many times. Then she turned on the light to look more carefully, but the light woke up the girl, who sat up in bed. Kamariya was looking inside the folds of the scarves. Then she felt around her neck and bosom. The little girl asked, 'What's the matter, Kamariya?'

Kamariya held the girl by her shoulders and asked her in English in a voice full of entreaty, 'Baby, you see my locket?'

The girl shook her head and joined Kamariya in her search.

By evening Kamariya had abandoned all hope of finding the locket. The desire to go home for Eid had all but died. It somehow seemed utterly meaningless to go home now, without her cherished asset—the greatest reward of her thrift and economy. By now everyone in the village would have come to know that not only had Kamariya again become prosperous but that she was also actively collecting things for her marriage. She imagined the shame of having to get off the bus at her village with a bare neck and was consumed by sorrow at the loss.

When, finally, Madame found out about the matter, it was as though an earthquake had rocked the house.

Who had access to the part of the house occupied by Kamariya? A Tamil milkman; but he would hand in the milk bottle from outside and leave. The man who delivered the bread, then? Well, he, too, never set foot inside the house and did business through the kitchen window. The maidservant who worked in the opposite house? Not she, either; she hadn't been seen around lately—and even if she had been around, she was a Malay Muslim like Kamariya, and Malay Muslims are known for their impeccable honesty.

That left only one person—

Minachi!

Although Madame was quite guarded about voicing her mistrust of Minachi, Kamariya, already skeptical about the Tamil girl, now felt confirmed in her suspicions. That Minachi was a Ceylonese Tamil and, what is worse, a non-Muslim, and a Hindu to boot, was already half the crime in Kamariya's eyes.

That petty theft plunged Minachi into the bottomless pit of ignominy.

The next morning Kamariya saw Minachi come in through the door at her usual time, but she didn't react at all.

Minachi parked her bicycle outside the kitchen in the shade of the wall and came in wiping her sweat.

Kamariya had thought that Minachi, afraid that she might be caught, wouldn't show up that day. Her showing up convinced Kamariya that she was a habitual thief.

The two exchanged a few words, after which Minachi picked up the dirty laundry and went into the bathroom.

A little later, Kamariya, too, came into the bathroom, and she began snooping around and rummaging. Minachi asked, 'What are you looking for?'

Kamariya fixed her gaze upon Minachi's face and answered, 'I've lost something.'

But Minachi's face didn't change colour, nor did her hands falter.

After some time she asked, 'Any luck!'

'No,' Kamariya replied on her way out of the bathroom. Then, suddenly, she volunteered: 'My locket and gold chain are missing since yesterday. Even Madame knows about it.'

Her last sentence sounded like a threat.

In the bathroom Minachi's hands went about their business, without the slightest pause.

After a brief interval Kamariya reappeared in the bathroom, once again riveted her gaze on Minachi's face, and said, 'But the master of the house still doesn't know about it.'

This was the second—and bigger—threat.

Her work done, Minachi silently left the house and without so much as looking at Kamariya picked up her bicycle and rode away.

Minachi's displeasure was beyond Kamariya's understanding; and the failure of the threat—'But the master still doesn't know about it'—to produce any effect meant that she had lost the necklace beyond all hope of recovery.

After this encounter Minachi plummeted so far in the esteem of the entire household that she had to be constantly watched by someone as long as she worked in the house. Madame herself, one of the boys or the girls, Kamariya, even the little girl—someone or the other always managed to be near her as long as she remained inside; and when she finally left the compound, a meticulous count would begin of the washing left to dry on the clothesline. The children would examine the banana, *kathal* and *chempedak* trees—who knew, she might have picked a few on her way out.

It was then that Madame suddenly remembered the blouse which had disappeared six months ago. It had certainly not been lost at the cleaners; rather it was lost somewhere at home, possibly in her own bedroom. The amazing thing, though, was that Minachi was not even allowed in Madame's bedroom. Often one also noticed a few fruits missing from the trees, and every now and then a few coconuts disappeared from the garage, where they had been stored away, but nobody paid any attention to that.

The incident of the theft of Kamariya's locket was a spotlight which fell directly on Minachi and made her stand apart from the other characters around her. But, interestingly enough, instead of appearing embarrassed, she seemed more lighthearted than ever.

Now after two days' silence, Minachi had asked Kamariya about the locket and, having been threatened with *bomoh,* the

Malay medicine man, and *Bachaan Yasin,* she had quietly wrapped her sari tightly around her legs, hopped on the saddle of her bicycle and ridden out of the compound.

All day long Kamariya remained in her room, and she cried intermittently. She mourned the loss of her necklace but deeply regretted having told lies during the month of fasting. That she had asked *bomoh* for help and that she had written to her father to hold the *Bachaan Yasin* were both untrue; she had merely used them as a threat. Then, there was a third 'sin'—falsely accusing someone—which was diminishing the reward of her fast.

At sundown, as she put a dried date in her mouth to break the fast, Kamariya ardently prayed for the recovery of her necklace—because prayer at such an auspicious time never failed, like *Bachaan Yasin,* to have effect—and later tried to absolve herself of the sin: 'After all,' she reasoned, 'I didn't openly accuse Minachi of stealing; my words, therefore, cannot be considered an accusation.'

It started to rain after *iftar* and continued throughout the night.

When she got up in the morning, the sky was covered with low hanging clouds, which meant that one could expect it to drizzle throughout the day.

Minachi arrived at work unusually early, her face free of the tension of the past few days. She left her bike outside against the window and rushed in, and said, in between gulps of breath, '*Kakak*, I bet you didn't sleep well last night?'

'On the contrary, *adek,* I slept very well,' Kamariya replied in a composed voice. 'What's lost is lost. Worry isn't going to bring it back. So why worry?'

Minachi sat down beside her on a stool, as though to catch her breath. Then, after a few moments, she asked, 'Can I help you search for the locket?'

Kamariya remained silent. The mental agony in which she had spent the last four days—the painful image of returning home without the necklace, the loss of an object bought with

hard-earned money, and the sin of falsely accusing a person—
all these seemed to have been washed away by that *iftar* prayer.

Minachi said impatiently, 'Last evening I went to the temple;
at night I had a dream in which the spirits of my ancestors
spoke to me. I could not make out their faces but could easily
identify each of them by their voices. They told me, Daughter,
don't worry. The locket and the chain have gone nowhere; they
are still in the house, somewhere in *amah*'s own room. So go
and help her find them."'

Kamariya's face, which up to this point revealed only
boredom, lit up with hope.

'*Kakak*, may I help you find it?' Minachi said, and without
waiting for the answer she began to walk to Kamariya's room.

Kamariya followed her into the room. A short while later the
little girl also joined the two in their 'game.'

They went meticulously through everything in the room. They
searched for the necklace in the table drawers and the folds of
every item of clothing. Once the little girl surprised them, with
'It's there!' And an impatient Kamariya swiftly turned to her
and asked earnestly, 'Where, baby?'

The girl pointed at the ventilator shaft close to the ceiling
and said, 'There! My doll! You see it, don't you?'

Both Kamariya and Minachi laughed. The girl had in fact
found her lost doll stuck in the grille of the ventilator shaft.

In the meantime, Minachi began to search through Kamariya's
bed. 'Right inside *amah*'s room—those voices told me,' she
said.

Kamariya was now beginning to have confidence in
Minachi's 'voices.' It seemed her earnest prayer at *iftar* was
being answered by the voices Minachi had heard in her dream
following her visit to the temple.

Just then something fell smack on the floor from under the
mattress. When the little girl crawled under the bed and emerged
with it, lo and behold, the necklace, which had kept Kamariya
sleepless for the past four or five days and had so preoccupied
her mind that she could not even single-mindedly devote herself
to worship, glittered in her hand.

Raindrops were dripping on the roof.

Kamariya was weeping silently. And Minachi, on her way out, observed, 'It's sopping wet today. The washing won't dry. If I wash it and leave it wet, it will start to smell in a day. So tell Madame that I will come and take care of it tomorrow.'

'All right, *adek,* I go tell Madame,' Kamariya said in English with mixed feelings of suspicion and joy.

Three days later Kamariya, daughter of Muhammad Yusuf, left for her village some 350 miles away in Kedah near Baling and surrounded by rice paddies and coconut palms.

Minachi was at best being tolerated at Madame's now. Everyone thought the dream incident was a mere hoax and Minachi's search for the missing necklace an elaborate ruse which she had herself contrived. After all, it was conceivable that she had somehow sneaked into the house quite early in the morning while everyone was still asleep and, after stashing the necklace under Kamariya's mattress, had left undetected. Likewise, it was also possible that she had the necklace in her hand all along. At the right moment, she had slipped her hand under the mattress and dropped the necklace on the floor with the skill of a juggler.

This was also Kamariya's opinion before leaving for her *kampong.* She had searched her bed many times. It just wasn't possible that the necklace had lain there undetected all this while.

At any rate, Minachi's presence in the house was now like a part of the body which could neither be severed nor put to any practical use. Each item was meticulously counted before she was given the washing and, once again, after she had left for the day; and someone had to keep an eye on her as long as she remained on the premises.

The tenth day after Eid, Kamariya returned from her village late in the evening. She was feeling rather tired from the long bus ride. At home, everyone gathered around her. The little girl came and sat in her lap. Kamariya patted the girl's cheeks and asked her mother in English, 'Baby no take much *susu* daily, Madame?'

'Plenty, Kamariya,' Madame replied lovingly.

Ignoring Madame's answer, Kamariya asked the girl, 'No *mandi* today, Baby?'

The girl wrapped herself around Kamariya and said, 'I do *mandi* with you now Kamariya.'

Everyone laughed at the girl's sudden indifference to everyone else. Without her *amah*, she had felt out of sorts all these days.

Kamariya opened her traveling bag, took out the gifts, and began distributing them: fruits such as *manggis* and *rambutan* were for all; the oldest girl got a necklace of oyster shells; the other two girls, each, a pair of straw slippers; the boy, a straw hat; and Madame and her husband, some other gifts.

Then, putting a packet back into the bag, Kamariya asked, 'Minachi come Madame?'

'Yes,' Madame replied, shaking her head. 'But I am thinking of firing her. Now that you are back, do look for another washing girl.'

'Why Madame?' Kamariya asked, astonished. 'I have brought a present for her, too.'

'Because she is a thief. She is not a good girl.'

'No, Madame, Minachi no thief. She is a good girl.' Kamariya's face was perfectly calm.

'Then who is the thief?' Madame asked.

Kamariya remained silent.

'Just who stole the locket?' Madame asked with bitterness.

'Minachi, of course,' Kamariya replied in a drained voice.

'And she is not a thief? She is a good girl?' Madame asked, thoroughly fed up. 'You are a strange girl yourself, Kamariya.'

Kamariya thought quietly for a while and then said in a halting voice, 'She did steal, but how can she become a bad girl by just one such act? She still has fear in her heart.'

She continued, 'I had neither asked *bomoh* nor written to my father for help. Then again, *Bachaan Yasin* works only when the thief is a Muslim and is aware that the 36th chapter of the Koran is being recited by a group of forty people. But Minachi is a Hindu; how do you expect the pious formula to have any

effect at all on a Hindu?' Kamariya broke into ringing laughter.
'When I told my brother about it, he laughed. I said to him, "If
Minachi were a Muslim and a Malay, I would have gotten you
married to her." He said, "If she were a Muslim and a Malay,
you would not have suspected her of theft in the first place." '

Madame laughed embarrassedly and said, 'But she is a thief.
We thought you would hate her for that.'

Kamariya fell into thought: Yes, why didn't she hate
Minachi? Or, maybe she had hated her, just for a while, and no
longer did. Why?

The little girl, huddled in her lap, was playing with her shirt
buttons, opening them and closing them again. Madame's older
son and daughter were tense with expectation waiting to hear
Kamariya's reply.

A gust of fresh air, bearing news of imminent rain, came in
through the window and left the curtains rustling in its wake.

When Kamariya tried to answer Madame's question, her mind
thought its own thoughts, and words which were not in her
mind spilled out from her mouth:

'It's true, Madame, people can be thieves, too; after all
Minachi is a person. It is perhaps less surprising that she stole,
rather it is more surprising that the mere threat of *bomoh* scared
her off. I have seen people who, in spite of knowing the whole
Koran by heart, still don't feel at all the fear of God.'

But the answer to the question, echoing in Kamariya's mind,
appeared—like the first stretch of land slowly emerging from a
flood tide—like a refreshing truth with which, she realized with
a pleasant surprise, she had been unfamiliar until now.

'Hatred is no different than love,' she said, 'you have it, and
then again, after a while, you don't.'

The words, 'Close the windows, Kamariya,' jogged her out
of her reverie.

Madame and children were shutting the doors and windows.
Kamariya heard Madame say, 'Kamariya, what you say is
beyond me.'

The next morning, when Minachi walked in, Madame
whispered to Kamariya, 'Keep an eye on her and give her clothes

to wash only after you have counted them first.' Kamariya looked at Madame in a strange way, as though feeling sorry for her lack of understanding, and said, 'Why, Madame? That will break her heart.'

Once again Kamariya surprised herself. The thought that had occurred to her mind as she uttered these words was so completely different: 'Madame, you no find ever no new thought in you?'

Then, releasing her arrested breath, she thought, 'Perhaps, Madame and I are made of different clays.'

—Translated by Muhammad Umar Memon

Night of Torment

I

We waited the whole day for mother's return, but she neither came back, nor sent us word that she was safe.

Nowadays none of us really has the guts to step out of the house. The stores close early in the evening, and it no longer surprises anyone to hear an occasional gunshot ringing out during the night. On bad nights, the old woman in the neighbouring house leans out of her balcony and screams in Persian, 'Madam! Madam! What's going on?' But upon hearing mother's assuring, 'Nothing, nothing at all,' she goes back into her room, leaving the street once again enveloped in silence.

The sky above is the same as ever, and at night the breeze still rushes through the alleys and lanes, as it has always done, but the smell of burning gas from the oil refinery is no longer in the air. In other words, the air has become pure. But nobody is happy with this purity. Some even claimed that the earlier, fume-filled, gas-smelling air was actually better than this deceptively clean air. Our lungs refuse to accept it even though we know that it's healthier for us, just as sometimes a sick man cannot persuade himself to swallow food even though he is hungry; he sends it back saying, 'I don't seem to be able to keep it down.' People say the real reason the air is clean is that the oil refinery has been shut down, and everybody is staying indoors. This peace, this quiet cleanliness—call it what you will—is only a passing phenomenon.

That day everybody in our family went without food. We had surmised that something unusual had happened to mother. None of her usual friends or visitors—including the rich sellers of antique carpets, the police officers, and the servicemen—had come by our house to inquire after her, even though we were sure they must have known she was in trouble.

Here everybody is for himself these days; some are surreptitiously transferring their money out to Switzerland, and some whose families have already gone abroad are busy making their own plans to go across the border. But our family has neither gold to smuggle out of the country nor money to pay for our journey abroad. Boats, dhows—all means of escape are beyond us. And, in any case, where can we go? Or, as mother puts it, who's going to take us in?

Actually, a number of changes have taken place in our household during the past couple of years. Sailors and aircraft mechanics no longer come by our house, and even the visits of the army men and businessmen have become scarce. The rent for the place we live in has gone up so high that mother often wishes there were some other person around to share it with.

There are four of us children in the family: two brothers and two sisters, but, unfortunately, both the sisters are younger than the brothers. If that hadn't been so, the numbers of visitors to our house wouldn't have declined as quickly as it has lately.

Nowadays we live in a two-room house. The visitors sit in the room which is more fully furnished and has on its walls large, gilt-framed pictures of the Caspian Sea and of female bathers on its beaches. These pictures have been there for years and have begun to look dull and common now. But to our new visitors they might still seem exciting.

This room, which may be called the salon, is used by us, the children, during the day. We jump over or lie down on the sofas. When the air is filled with the fragrance of flowers, or with the smell of perfumes, we open the windows. The curtains flap gently in the breeze and make tiny waves.

The other room is the living area for us brothers and sisters. There are beds in it and a table and a chair. On the wall hangs a poster-sized picture of Caliph Ali, as big as the largest picture in the adjoining room, though printed on inexpensive paper.

This back room is airless, and even the furniture in it is of the ordinary kind. It is painful for us to recall that all the expensive pieces of furniture have gradually been sold during the last few years.

Until recently, at night this whole neighbourhood used to glitter with bright lights. As soon as it was dusk, the lights came on everywhere, and women began to appear in their windows. Cupping their chins in their hands and resting their elbows on the windowsills, they sat as if merely watching the world below go by. Some women or girls sat casually astride their doorsteps, their backs leaning against the door frames.

At times, a sailor passing through the lane below would stop and with his new Japanese camera take a photograph of one of the girls. The light of the flash bulb would light up the window, as lightning suddenly brightens the sky. The photographed girl often smiled and waved her hand, and sometimes said a word like 'Ciao.' A long time ago we saw an illustration in a book— three children with their older sister, sporting a red scarf around her neck, peering out of a rough-hewn window frame. That illustration must have looked like some of those occasional photographs taken by the sailors. But none of those passing sailors who took mother's picture from the street below and who later on, upon being invited, came upstairs, ever sent us a copy of any of those pictures. But such incidents do not bother mother. She has always smiled for everybody's picture, just as little girls do.

The day passed and then it was evening. Then slowly the night began to fall. It was quiet all around, and the silence began to gnaw at us. We had no money at all, not even for a Pepsi and a mutton-kebab sandwich—mutton because that is the cheapest of all meats; a small piece of mutton is strung on a skewer, and then a piece of onion or a wedge of tomato is added, then another piece of mutton and some more pieces of onions or wedges of tomatoes are added, and so on. In all, the whole skewer holds only three or four mutton chunks. And although mutton has an unpleasant odour to it—unlike the kebabs from good restaurants, which have an appetizing aroma—when we were hungry, we would eat it without quibbling or fussing over the smell. Mother has always admired our good manners in this regard. Whenever a guest sat—actually, used to sit would be a more accurate expression—in the adjoining room (the room with the pictures of the Caspian) to drink his whiskey and sent for some *chilo* kebabs and pilaff for himself and mother, she would, on some pretext, often bring her plate to our room and leave more than half her share for us to enjoy, whispering as she left it, 'Don't fight over it.'

That night we didn't get even a wink of sleep. Our toes began to hurt from standing as we peered out of the window. The back lane, where once women wearing full make-up had often sauntered about in the darkness till late at night, or where they had stood in half-lit doors lifting their blouses and blowing on their breasts, was now like a graveyard. On the main road a military vehicle would hurry by every hour or two.

Mercifully, there were no sounds of gunfire that night. Our stomachs were empty. Our hearts were going pit-pat, and we were worried sick over what mother might be going through. Our eyes smarted from the lack of sleep and our legs were in constant pain.

But somehow that night passed. At dawn we saw mother staggering into our alley after turning the corner in front. Seeing her in such an utterly pitiable state, the two older children just stood there, dumbfounded. The older girl let loose a loud scream, and the younger one fainted.

We don't know how mother had the strength to walk to the door of the building. She was like a toy operating on a nearly dead battery or on a nearly wound-down spring. She fell down in a heap as soon as she reached the door.

We ran down the stairs. A few women also came out of their apartments. They revived mother by splashing water on her face, and when she groaned, they tried to help her stand up. But when she couldn't, they just lifted her and carried her upstairs. We, the children, followed them quietly, as if walking behind a hearse. They put mother down on the big sofa in the salon.

One woman asked, 'Is there any brandy or whiskey in the house ?'

'How could there be?' the other answered bitterly. 'It's futile even to look for it. Whatever was left with anybody, those cloak-and-rosary men took it away in the raid the day before yesterday.'

The old woman from the neighbouring flat who sat there rubbing a wet handkerchief gently and sympathetically on mother's lips pulled herself up with difficulty. To stand up she had to push her knees with her hands, as she was known to suffer from arthritis. Everybody made way for her to pass. A while later when she returned from her flat, she was holding in her hand a small perfume bottle wrapped in a handkerchief. Knowing smiles appeared on the women's faces. Perhaps taking her to be an old hag, no one had bothered to raid her flat.

The old woman sat down once again near mother, on the carpet. She mixed a few spoonfuls of brandy in the glass of water which had stood untouched on a table near mother and asked her to drink it. Mother came out of her stupor as soon as she had the brandy. She tried to get up and walk towards us but only staggered and fell back down heavily. We ran to her and hugged her. She held us close and began crying and sobbing. Tears came to the eyes of many who heard her lament. The women gradually began to retire and go home.

Mother didn't seem much concerned about the lash marks on her back. Her skin had split in many places where the whip had struck her and in some places her shirt was clinging to her skin because of the dried blood. The women were trying to loosen it with water. Some lash marks had even come up to her breasts, as though the whip were a serpent that had coiled around her and reached in front after having struck her repeatedly on the back.

Whispering to each other the women finally departed. They had been so concerned about mother's lamentation that none of them had even bothered to inquire if we had had anything to eat all day yesterday or last night.

Yesterday when the 'cloak-and-rosary men'—the members of the Government's new morals patrol—had raided our house and taken away, in their police van, mother and her visitor, a man who was from out of town, the women who lived in the building had shut the doors and windows of their houses. Every door and window in the building, except the window of our house, had remained shut, and terror had reigned over the city.

Now, with mother's return, everybody—for the moment—breathed freely again, at least until such time as another woman would be taken in the police van to some unknown place, to return home twenty-four hours later, in a similar condition, staggering and half dead.

II

Most of the questions I was asked were beyond my ability to understand or answer—for example the questions about the *Shariat* laws. I wanted to tell my questioners that I had never been taught about the *Shariat*, nor were there any schools, or academies, or even libraries where I grew up. Learned and religious men did come to our area, but only for brief visits, never to stay there for long. Some who came from foreign lands and were ignorant of our language often inquired about 'the way up' in words which even illiterates like me knew had been learned and memorized by those people in their childhood in religious schools.

But in these troubled times, how ridiculous must our plight now seem to the same foreigners who would be sitting comfortably and peacefully in their homes, laughing at us. But there was one thing quite curious about the district I lived in: no matter when it was built up, it was meant for those who, of necessity, had to be temporary residents. Strangely enough, its population had never decreased, but only became bigger and bigger. So then where did those girls come from who were there day after day? Had they fallen from the sky?

But I didn't open my mouth to say anything, for women like me do not have any practice in the art of oratory.

This place where I was—whether a police station, a military barrack, a big hall in a former palace—whatever it was, was steeped in silence. It was the kind of silence which made every sound resonate—like the thud of the heavy army boots, or the whiz of an automobile passing by in the distance. It was a silence that made each sound carry an impression of its own, separate and distinct from other sounds, not as part of a group of sounds. In front of me hung the new slogans of the revolution, painted on white broadcloth which had been nailed to the wall. Red dust which had fallen from the wall due to the crude handling and nailing of the slogans had streaked the white cloth. It looked as if the nails had been driven in very clumsily. Here and there one could see small no smoking signs written on pieces of cardboard which rested on wooden stands. The typist sitting facing the members of the tribunal was preparing a copy of the proceedings of the previous case. He had no interest whatsoever in my case, nor in the victims of the judgment rendered in the case he was busy typing up. The revolution seemed to have dehumanized him, transforming him into a machine. Nonetheless, earlier, when a bearded military officer had asked me my name and I had answered 'Fatima,' the typist cast a quizzical look at me, but then again became busy with his thumping on the typewriter.

My mind, when I faced the tribunal, began to wander. I didn't feel the least bit concerned about my erstwhile visitor who stood in one corner of the courtroom—or whatever that place was—terror-stricken, as though he had just come upon a ghost. Nor did I think about my four children I had been forced to leave behind, unfed at breakfast time.

Dozens of times I had asked myself the question: Do I have a will of my own, or did God, while He was shaping and forming me, forget to give me an individual will?

Apparently, neither I nor my children—that horde descended from the sky—had any claim to any place under the sun; also, it seemed that nobody, not even God, was willing to take on the responsibility of feeding us. 'Not even God'—that nagging phrase would raise its head again and again in my mind, while I tried to push it back down each time. I had guessed what the verdict of the court was going to be in my case, having overheard the whispers of the functionaries who brought me in there. My mind had already gone half numb, and that phrase, 'Not even God,' despite my effort to suppress it, had begun to echo again and again somewhere in my mind. It acquired an identity of its own, like the identity silence gives to sounds, as I have said above.

In the district where I live, the faces of girls and women look exactly alike at night.

Girls clad in colourful garments, their silky brown hair cascading down their necks and bare shoulders, their faces yet unmarked by lines of age or care, would doll themselves up every evening and take their places in the windows, looking very much like the pictures in the books our foreign visitors who stayed overnight with us often brought with them to while away the time. I know a smattering of German and French, as well as some English, and I've often seen my young visitors laughingly wave those pictures in front of our faces and say, 'Like you, eh?' meaning that the girl in the picture looked like me or one of the other girls there. Once when a young man waved a page from a book printed simultaneously in four languages before my eyes and said, 'Omar Kheyam?' his companion cut him short and said partly to him and partly to me: 'All girls here are like Omar Kheyam. See! Even you look like Omar Kheyam to him!'

At that time, as I stood facing the tribunal, I remembered those pictures in the books, pictures of young girls with their curly hair, their eyebrows raised at the edges, looking like birds about to take wing, lending the girls' faces a certain dignity and haughtiness, their breasts sculpted from the whitest marble and visible through their see-through garments. The pictures also showed faces of bearded men, both young and old, holding goblets of wine in their hands, looking beseechingly at the girls, as if asking for something—life? escape from life? deliverance? Who knows. The girls in those pictures all looked alike.

The same bearded faces of the men in the pictures were also there in front of me at that time; some of the men were religiously picking their teeth after lunch, while others, oblivious to us, were whispering into each other's ears.

For me, though, that hour of the day was the hour when every woman in my district had slipped back into her own face, strained and tired, her fading penciled eyebrows revealing the stubble growing underneath on the wrinkled skin. At that hour of the day even the passersby were different from those who strolled down our street in the evening. There were no wealthy businessmen, no dealers in gold or opium or rich carpets. These noblemen were replaced by the riff-raff, the laundry men, the vegetable vendors and others whose job was to look after others' material needs.

For some time now even these lowly types had become scarce in our lanes. A number of women from our neighbourhood had disappeared, forcibly abducted, never to reappear. To begin with, these women very seldom had any men in their houses, men who could ask the authorities about their whereabouts. And if there were any, they wouldn't step out of their houses for fear of being shot.

The first question they would probably be asked: 'Do you share in the profits made by that "fairy" of yours?' And even if they could come up with an exonerating argument, an irrefutable alibi (such as, 'No, that was the way things have always been'), since the government didn't have any other plans for such people, nor yet another world-view, these men would be taken away even while they were speaking in their defence and lined up against a wall (perhaps somewhere in the courtyard of the same building where I was); they would hear the heavy thud of army boots, the sound of American rifles being loaded, and some meaningless words (of the policemen, of the officers commanding them, pleas for mercy, a recitation of God's names), and then a loud report, followed by a long silence.

My older son has a Christian friend whose father drives a taxi in town. A picture of Mary, mother of Christ, holding the infant Jesus in her lap adorns the dashboard of the taxi. I've noticed that picture the few times I've travelled in that taxi. I've seen the driver make the sign of the cross when he narrowly escapes an accident. My son told me the other day that within the last few weeks his friend had become a totally changed person. He told my son that his family had become completely disenchanted with the country and that they were planning to emigrate to some other country because, in his father's view, their fellow citizens, the non-Christians, had suddenly become unrecognizable as human beings. They were no longer the people they used to be. As a result, he had been assailed by all kinds of fears—the fear of being publicly flogged, of one day being pushed up against a wall and shot. His greatest fear was that his own children might come to be coloured by the world around them. For centuries these people had lived in this land, never fearing that their children would give up Christianity and convert. For them there was nothing to fear from the religion of the majority until now. But now he was scared of the outward form that that religion had taken, the form which threatened to teach his offspring not religion but a creed of violence and tyranny.

III

Come you spirits
…unsex me here,
And fill me from the
      crown to the toe top full
Of direst cruelty.
—William Shakespeare, Macbeth.

By evening my erstwhile guest was in pitiful shape, while I sat on a bench empty-handed. We hadn't been given anything to eat or drink since morning.

The people who made up the tribunal kept changing during the day. Most of them had beards. These were the people known these days as 'the men of the cloak and the rosary.' The bench I sat on was at the back of the hall. From there I could observe everything, see everyone who came in or went out. Sometimes a person would rush in, hurry by the dais on which the members of the tribunal sat, and bow down respectfully with a hand on his chest before rushing out the opposite door. Once or twice during the day we heard shots being fired. The sound made my guest shudder. He and I sat quite far from each other, or I might have spent that day of hunger and thirst talking to him.

Many times I tried to catch the attention of anyone among those who passed by me, but it seemed they were not human beings but rather mechanical appliances devoid of all feeling. One could even have called them religious robots.

At last, at the time when the darkness of the evening had begun to show through the windows, and I had just about fainted with hunger, someone shook me and asked me to come forward. It was now our turn before the tribunal.

My companion cried, 'No, no! For God's sake ...,' but the soldiers once again commanded him to move forward.

I was surprised to notice that in that room where only men had come and gone the whole day, a woman had quietly entered—I don't know when—and now stood respectfully on the left side of the dais.

I was in a sort of stupor, all my earlier fears having been drained out of me during the passage of the day.

A cloak-and-rosary man asked me the same questions I had been asked in the morning.

'Do you know the punishment for this crime of yours?'

I kept quiet.

He asked my companion the same question. He couldn't remain silent but began uttering, without really any reason or hope, desperate appeals for mercy.

The cloak-and-rosary men present there in the evening were perhaps not the same ones who had questioned me in the morning, or perhaps they were the same, but having gone through a whole day of issuing and handing down punishments, they had lost count of who and how many they had punished and had, thus, become oblivious of our very existence in that hall.

The woman standing in front of me was about my own age, but she had donned a special type of uniform. I wondered if she had any children, and if she did, whether she had, like me, left them at home that morning, untended, to come there. More than in my erstwhile guest, I felt interested in her at that moment; in fact, I felt as if I were being drawn towards her by a certain kinship. Maybe, I thought, her husband was home looking after the children.

I was asked: 'What is your means of livelihood?'

Respectfully I answered, 'You know it already.'

One member of the tribunal looked at me with glaring eyes and warned: 'Don't forget that you are standing before a court, and anything you say might go in your favour or against you.'

I nodded.

Another member of the tribunal asked: 'Who fends for you?'

'I do it for myself.'

'Does anyone else help you in this? I mean, who is your supposed husband?'

I kept quiet for a while and then answered, 'I have four children.'

'We know about all that,' one of them barked at me.

Suddenly I was gripped by the fear that they might have been to my house in my absence. Greatly disturbed, I asked them that question.

My question pleased the members of the tribunal. They all smiled.

Then one of them looked me in the eye and asked: 'Who is their father?'

Another one rephrased the question, as if correcting an error: 'Who are their fathers?'

For a moment I was tempted to name the real fathers of all my children. But would that have served any purpose? The fathers of my children were still tied to me by an invisible string and it was within my capability to drag them into that court at that time. But I did not want anyone else to go through the agony I was going through.

'They have all fallen from the sky,' I finally said.

The members of the tribunal admonished me harshly to mind my words. In their view, I was already past any moral good, having reached the lowest depth of degradation. I had not just lost my virtue but had even become crude and unfeminine.

Within a couple of minutes of the last comment they read out their verdict. Citing chapter and verse and referring to religious decrees, they awarded me forty lashes and my erstwhile guest fifty. And the money that was found on my person was confiscated as 'wages of sin.'

Hearing the judgment, I swallowed with difficulty the thick spittle that remained in my mouth, but my companion fell to the floor in a swoon.

The two soldiers walked to the back of the hall where I had spent most of the day, picked up the bench and brought it to where I stood at that moment. They did the same with the bench on which my companion had spent a tearful day.

Then, upon a command from a cloak-and-rosary man, the woman approached me and asked me to lie down on my stomach on the bench.

Her job seemed peculiar for a woman. I looked her in the eye and asked, 'So, it's you who'll ...?'

'Yes,' she answered.

I wanted to talk to her, to ask her if she had any children, but by that time my hands and feet had been tied with a rope and secured to the legs of the bench. The woman, devoid of human feelings, a cog in that huge religious machine, was standing over me, on my left side, holding a whip in her hand.

My punishment over, I was untied. My companion too had gone through his punishment and lay unconscious on his bench.

* * *

A few hours later when I could see the early morning light spreading through the windows, I was let go with some advice on leading a chaste and pious life.

'Did you realize the nature of your crime?'

'No,' I said.

Annoyed and angry, the cloak-and-rosary man said: 'The proper punishment for the likes of you is death. You have lost all sense of guilt or shame.'

I said, 'My lord, a lot more besides the sense of guilt died in me today. But if you really want to know, I never had any sense of guilt.'

He raised his hand to slap me, but I addressed him with courage—the kind of courage that wells up in those who are at the brink of extinction. I said, 'I feel sorry for you.'

He held back his raised hand and asked: 'For me? Why?'

'For what you are doing,' I told him.

'What do you mean?' he asked.

'I mean that just as men have changed the course of my life, never allowing me to become what I could have, or what any woman could have, in the same way you have brutalized this other woman as well. She should have been rocking a cradle and singing lullabies, but just as you purchased me, you have purchased her as well, and put a leather whip in her hands.'

* * *

If you have the leisure
Consider it a blessing,
That you enjoy the company
Of a cup-bearer or singer,
Or of song or wine;
Forsake those forever who cheat
Their God with prostrations
And their Prophet with praises.

                              —Asadullah Khan Ghalib

—*Translated by Faruq Hassan*

A Tough Journey

There was really no good reason to take the *pattewala* along on the trip. His place was in the office, outside the boss's room, sitting on a stool or a chair, the water jug nearby, covered with tiny flecks of mold, a cruddy steel tumbler resting on top, dented from countless falls, knocked completely out of round.

But if there wasn't much reason to take the *pattewala* along, there was even less reason for him to *go* along. The boss just happened to mention it—and not even to the *pattewala*, at that: 'I'm leaving at 6 am tomorrow. Make sure the jeep is okay. Water, petrol, battery, tire pressure, spare tire, everything.'

At this the backup driver asked, 'You going alone?'

The boss wagged his head yes, since the permanent driver was on a six-month holiday.

But the *pattewala*, who at the moment was pouring water from the filthy jug into the filthy tumbler to give to some visitor, was close enough to the boss's room to overhear every word.

The backup driver had returned to work just a few days earlier after a spell in the local hospital's bone and joint ward, and even now some of his body parts were still covered with white plaster. His hand should have been immobilized too, but he was as reckless with it as he was with the other, plaster-casted parts of his body. Indeed, as reckless as one presumes he was with the government's vehicles.

In any event, as soon as the backup driver came out of the boss's room, the *pattewala*, without even waiting to see if the visitor would tip him for the water, marched straight into the boss's room and said in a voice at once toadying and decisive, 'Chief, I'm going with you.'

'Where?' the boss asked.

'Where are you going?' the *pattewala* asked.

'Goth Haji Sadiq.'

'Then I'm going to Goth Haji Sadiq, too.'

Collecting some things from his desk the boss said, 'There's nothing to see there. You sit tight here. There might be a phone call or something. Maybe someone'll come while I'm gone.'

But the next morning, when nine o'clock came around and the boss was still trying to get out of bed, the gardener came inside and said, 'Boss, the *pattewala* wants to know what time you want to leave.'

Lifting his alcohol-soaked head from the pillow, he asked, 'Who wants to know?'

'The *pattewala*,' the gardener said standing in the doorway.

'When did he come?'

'Beats me. He was sleeping on the grass when I got to work. Over by the papaya trees.'

Rolling over to sit on the edge of his bed the boss said, 'So what's he doing now? Did he go back to sleep?'

'No, Boss,' the gardener smiled. 'He's asking to have a "coop" of tea and a few "bishcuits" sent out to him. Maybe he didn't get his breakfast before he left.'

For a while it was silent in the room—in the whole house, for that matter, since the boss's wife was off to her village for a few days, perhaps for the circumcision of some relative's son, which the boss had absolutely no desire to see. As usual, he had no desire to see his own village either. These days at any rate, the only place that appealed to him was here. This place right here, where the twentieth century was grandly preparing for its hundredth anniversary. Against this, what did the village have to offer?—Drainage ditches full of filth; open sewers between the houses; snotty-nosed kids in the lanes; reclining old men spitting on walls; their young wives; flies; old junk behind the houses; quick, aesthetically tasteless romantic encounters in the fields; conformity; and a run-down 'reshtaurant', which was really just a tea stall where everything connected with it seemed, even to the proprietor and the tea-boy, to be smoked like well-dried fish. Even the cobwebs hanging from the ceiling had long ago turned black with grimy sediment. In the village it was as though a man married each thing there with no chance of divorce. Whether the village headman, some all-around good guy, the *imam* from the mosque, or a wealthy landlord, he just had to live with it, no matter what. All one's life, freedom was utterly impossible.

But here? Here there was every kind of freedom, a perpetually enduring newness. A man could unload his double bed on the junk collector when it gets old and buy a new one. He could change his neighbours. Sometimes he might pick one club or restaurant to be his drawing room, sometimes another. In the city it was all about choice: friendships and trusts could be changed, and if a man didn't particularly like his office or department, he could even get his supervisor changed. What benefit could there be in serving under some boss who's of no use to you?

The boss felt his heart catch. 'But this too is here: that magnifient sonofabitch, lounging around for a couple of hours on his boss's lawn, calling inside for someone to bring him some tea. Put him in the village, and see how fast he'd discover his proper place there! Instead of calling to have some tea brought from the house, he would have come bearing a proper gift himself. But here? Here he tells the gardener to fetch him a "coop" of tea and some "bishcuits." Doesn't even need to ask my permission! Well, I've had enough. What good is he? The babbling *sheedi.*' He hated the *pattewala*'s fat nostrils. He hated the way he talked, the way he walked, the oil he put in his hair, the puffs of cigarette smoke that came from his mouth. He hated the man entirely.

Slowly he turned his head to where the gardener was standing, and remaining blankly just so for a while, he finally said, 'Okay, tell Joriyo to take some tea out to him.'

'And the "bishcuits"?' the gardener asked.

'Those too.'

'But Boss, we don't have any "bishcuits",' the gardener said with a devilish grin.

'This one too is a first-rate sonofabitch,' the boss thought to himself. 'How the hell should he know if there are any biscuits in the house or not? How come I never hear him saying, 'Boss,

I have to get some seeds,' or 'Boss, we need some more fertilizer'?'

But then he recalled that the gardener had indeed said just this many times before. But only when his (the boss's) wife would say, '*Mali*, just look at the condition you've left the garden in!'

Joining his hands reverentially the gardener would reply, 'Begum Sahib, the soil's begging for fertilizer. There's none at all in it. We've been out for a long time. And I've already asked you about the seeds, too.'

The boss found himself thinking, 'In the end, what good is this gardener?' But the words from his mouth were these: 'Send him some of whatever there is—a *paraathaa*, a piece of cake, some toast. And tell him to go back to the office when he's done with his tea.'

* * *

So this was how the *pattewala* and the boss got together.

The road was deserted in front of them, and dust was whirling about here and there. There was just an occasional breath of fresh morning breeze left, the rest of the winds having already been made warm by the sun.

At first the *pattewala* sat in the back seat, but when the sand and rough road started to get the better of him he said, 'Boss, I'm coming up front.'

'Why?'

'The dust is really bothering me.'

And before he could be stopped—an action for which a good reason would have been required, which the boss's alcohol-soaked head, accostomed to thinking less even than this, would have been hard-pressed to produce—the *pattewala* scooted between the two seats and into the front.

On top of his well-oiled head sat a cap covered with small sewn-in mirrors and decorative piping. His clothing was of the traditional, local variety, made of some nice shiny fabric, over which, in this heat, he had donned a heavy nylon waistcoat.

'Maybe he imagines we're going to some wealthy landlord's feast,' the boss thought. But the *pattewala* was thinking something else. The boss was about forty-five, the age when people start to become—if not paternal, then at least maternal—grandparents.

The *pattewala*'s nostrils were shining in the sun, as was the watch strapped to his wrist. It was hefty, just like its owner. How strange: he shines in the sun because he's so dark, but his watch because it's so bright and silvery.

The boss's watch was slender and golden, the features on its dial quite delicate.

Physically too, they were as different as night and day. The boss was the kind of guy who could eat anything and not show it. A small paunch had begun to emerge, but this was only after years of drinking beer. He had been drinking hard liquor regularly for just a few years, however. A scotch or two at night while watching the V.C.R., more if he went to dinner somewhere. As his years of service increased, so too did the number of dinner invitations. The relationship between drinking and thinking defies logic. So too between the number of dinners and the capacity for caution. Accordingly, even now he remained silent, inside as well as out.

The *pattewala* took a pack of cigarettes from his pocket and offered one to the boss. Seeing him shake his head no, the *pattewala* carelessly lit one up for himself, took a few drags, and opened and closed the cover of his Ronson lighter a few times, pleasing himself simply with the satisfying sound of its 'ka-ckick.' Looking ahead into the blowing dust, he said, 'Boss, I'm not being treated fairly.'

The boss remained silent. This one sentence of the *pattewala*'s, coupled with the boss's silence, only made worse the tension between them.

The jeep crossed a canal in which two men were washing their horses. On the bank some more men had wrapped their shoulder-cloths into rings around themselves; they were squatting inside them, supported on their haunches like birds resting on shallow nests in the noontime sun.

The *pattewala*'s presence was weighing heavily on the boss.
A weight of unease, of agitation. What the hell could be going
on in the *pattewala*'s head? What kind of schemes, now that he
had an entire day and night away from the office—and with it,
total freedom?

Word had come from Goth Haji Sadiq that the irrigation tank
there was completed, and if everything was okay, i.e., if it could
hold water for even twenty-four hours without leaking, then a
week later they could have the Minister for Rural Development
or his military attaché perform the opening ceremony, a day or
two before Ramzan. Regardless of whether this ceremony would
have any effect on the villagers or not, the newspaper, television,
and radio coverage would ensure an impact outside the village.
Ramzan was coming up quickly, and it was like the publicity
officer Mubashshir had said: If we don't print any pictures of
the people getting water from the tank on the day of the
inauguration, if we hang on to them, then a week or so later we
can twist the caption. For instance, 'A CROWD OF FASTERS BEFORE
SUNSET AT THE NEW IRRIGATION TANK.'

They could also run an interview with one or two of the old-
timers. Something like: In the olden days our women had to
trudge two miles for water, but now the tank has become a huge
convenience. We perform our ablutions there, we use its water
for drinking and cooking our food. So we're thankful to the
government from the bottoms of our hearts.

All the same, the arrival of the military attaché was a bigger
event than that of the minister, and once he was gone there
would be no one who'd even ask about the tank. The full weight
of the water could crack the tank and it could all run out, it
could seep from the walls and floor and disappear—no one
would care. Once the ceremony was over, the boss, who was
the project's engineer, wouldn't give a damn, even about where
the water was supposed to come from later on.

For a time, the canal a quarter-mile beyond the tank had been
running full. Water was diverted from it to the tank, and the
canal must have turned bone dry for the whole week afterward.
Indeed, it was a canal really in name only. Constantly losing its

water along the way it became so small just beyond Goth Haji Sadiq that the children could only splash around in it. The boys made a game of leaping from one side of the stream to the other. Provided, that is, there was any water at all. The tank had been constructed for just this reason, right where the rains formed a natural reservoir. But it didn't rain every year, and this structure of steel beams, cement and tar, despite its beauty, was both smaller and shallower than the natural reservoir. Everyone knew that the people of Goth were grumbling about how small it was.

'Boss, the Goth people aren't very happy with the tank,' the *pattewala* said, letting loose a fog of smoke on a butterfly that had somehow flown inside the jeep and landed on the windshield.

'Yeah, so you know I'm going to inspect the tank,' the boss said shooing the butterfly off the windshield in front of him.

The *pattewala* laughed the smug laugh of the experienced, especially as they laugh at the novice. 'Boss, it's pretty obvious, really. Next week is the inauguration ceremony. At the fort I've already seen the marble plaque they're going to put on it, too.' His tone of voice had changed in an instant.

The boss stopped the car under a tree, and lighting a cigarette for himself he asked, 'What does the plaque say?'

'Something in Arabic, probably a verse from the Koran. And then there's some English, which must be all the names and whatnot. The rest is probably the same stuff they put on the stone plaques for bridges and schools and such. I don't get much fun out of reading these plaques, and there really isn't any need, either. On top of this, the people from our department are first-class sons of bitches. They didn't treat you fairly, either.'

'How is that?'

'Down at the bottom of the plaque, where the construction engineer's name goes, they wrote some *hadith* instead.'

Leaning his head back against the seat the boss said, 'What were you doing at the fort?'

'My sister lives over there.'

'I bet she does. What were you doing at the engraver's shop?' The *sheedi* again laughed that same laugh. 'You just leave that one alone, Boss. There are a thousand reasons a man could go over there. Maybe my father-in-law died, or maybe someone's mother-in-law was dying. Stick to the issue and tell me this: how long am I going to have to put up with injustice?'

The shade under the acacia and *sheesham* trees was very pleasant, and Goth Haji Sadiq was not very far off. The boss said lazily, 'What kind of injustice?'

'Look Boss, I've been working longer than you've been alive. I don't know how many bosses I've seen come and go. I've lost track of how many clerk-types I've seen turn over. Someone has a mansion built, someone else does business on the Irani border. Even every L.D.C. and U.D.C. gets some respect. But me? Hell, even my woman doesn't talk straight with me.'

'You like drinking, don't you,' the boss said.

'I swear!' said the *pattewala,* getting a little angry. 'If I've gotten drunk even once in my entire life, then my father, and his father, and *his* father, all the way back seven generations— you can call them all a bunch of drunks! I'm not kidding, seven generations! I know what I'm talking about. Okay, I'm a tea addict, number one, and cigarettes number two. But only this brand. I swear, I don't even touch other brands.'

The boss took a glance at the *pattewala*'s pack of cigarettes and said to himself, 'This sonofabitch is a real bastard.' He went to start the jeep.

The *pattewala* said, 'Let's just stay right here for a little while. We can talk about things. That's why I came along.'

'What 'things'?' the boss asked, a little irritated.

'Just what I've been saying. Have I been treated unfairly to this day or not?'

A truck loaded to the sky with chaff passed by, shaking the jeep and sending bits of chaff flying everywhere, even inside the jeep.

Brushing off his pants the boss said, 'But you don't say *how* you've been treated unfairly. You've been beating around the bush for an hour: a headstone for your father-in-law's grave, your wife doesn't talk straight to you, people are doing business on the Irani border. Get to the point or sit and be quiet.'

'You want me to get to the point?' the *pattewala* said.

'Yes.'

'Do you really?'

'I don't have to put up with this nonsense!' the boss said in some heat.

The *pattewala* took no notice of the boss's anger. 'Just don't blame me later on. I'm not in the habit of talking a lot. I'm telling you because you said to.'

The boss remained silent.

'Here's justice for you! Big deals being cut, the commission keeps going up and down. Yesterday some guy comes with hands joined and *salaams* me seven times just to get me to take his application in to the boss. Today, after he gets his nice desk and chair, he's completely transformed. But me? I'm the guy who just sits, keeping watch like a dog at the door for twenty-five or thirty years.'

The boss said, 'So why don't you quit?'

'Sitting on that *pattewala*'s stool all these years, cutting off circulation so my ass has turned white, where would I go if I quit? The little respect I do get when each visitor greets me, then asks me something or other.... You know that too would be finished.'

'Then what's the use of crying over it?' the boss said reaching for the jeep's ignition.

'Wait a minute, Boss. We'll get to the point right here.'

'I don't have to discuss any of your "things" with you, I don't have to help you get to any "point"! Are you my equal, that I should 'get to the point' with you?!' the boss said trembling with rage.

'Don't get angry, Boss. You're like my parents to me! I'm your son, your *put-vaangoon*! Insult you? *Saa'in munjhna!*' he said touching his earlobes in penitence, drawing his index fingers down along his nose. 'I couldn't even think of it!'

If the words a weak man speaks in anger are taken to heart, his trembling gradually subsides, and not only does his chest swell with pride instead, but his face too assumes a look of gravity. With that gravity in his voice the boss said, 'Okay, forget about the jeep for now. What do you mean by 'getting to the point right here'?'

Offering the boss his pack of cigarettes the *pattewala* said, 'It's to your advantage, too. Here, have a smoke.' The boss's hands remained on the steering wheel.

'*Sahab munjhna*, it's your brand, here, have one. And the commission: has it or hasn't it gone from eighteen to twenty-five?'

The boss looked at the *pattewala*'s face; it was the face of a negro. Having become so used to putting up with everything, it was only with difficulty that his face ever registered any emotion.

The boss took a cigarette from the *sheedi*'s pack and said, 'What if I tell you I won't discuss it?'

'But you've already said you would.'

'And if I tell you to get out right here?'

'First of all, you won't. Why? Because I know you're an honorable man. But suppose I am forced to—there are a hundred ways a man can be forced—even if I am forced to get out here, still, Goth Haji Sadiq isn't exactly a hundred miles away. You'll get there in twenty minutes, it'll take me an hour. Maybe some kind soul will give me a lift and I'll show up right along with you.'

The boss silently cursed the *pattewala* and smoked his cigarette. The *pattewala* took a handkerchief from his pocket and started cleaning the windshield from inside the jeep.

After a difficult silence the boss said, 'So what is it?'

'You want me to say?'

'Go ahead.'

Lighting a fresh cigarette with the end of the old one the *pattewala* said, 'Take this very tank, for instance. Some minister's coming to inaugurate it. Are the villagers going to see any benefit? God only knows. But top to bottom, *lots* of people have benefitted. You know it, and I know it.'

'Tell me, who benefits?' the boss asked, feeling the *pattewala* out.

'I know all about everything, Chief. You know that stool I sit on, right? You better believe, it's like sitting in the sky. I see everything. I don't have to do a thing, information comes by itself. So tell me. Where did the money come from? From the villagers' own pockets, that's where! The poor people, they were afraid. Their livestock were dying of thirst, their crops were already drying up. Somebody donated a rupee, somebody else a half, another a quarter, and they came up with the money themselves. But who got all the recognition, all the credit?'

'You seem to know the answer. Tell *me*.'

'The bo-o-o-o-o-o-ard! Those low-born sons of ...'

'Watch your mouth,' the boss cut him off.

Bending his thick features into a slight smile the *pattewala* said, 'How can we have a heart to heart talk if I can't swear a little? I know how to talk polite, but that's for the office. Listen to what I say, not how I say it. The tank is just like all the other business I've seen over the last twenty or thirty years.' He put his hand on his heart. 'It's all buried in here.

'The board doubled their profits. They got their money from the people, and they got their money from the budget. What they didn't eat themselves they slipped into the government's pocket. So tell me, do I know all about everything or not?'

The boss was helpless before this *sheedi*, like a well-to-do man who has to spend a night in jail. Giving not a thought to his prisoner's social standing, the two-bit guard stretches out his grubby feet right in front of the cell and dozes off into peaceful sleep.

'You sure do,' the boss said, eyes shut tight.

'But how much did the people give? How many feet of cement did it take? How much steel? Goth needs something adequate for all the people who live there. But how big a tank did that sonofabitch contractor build? No bigger than his father's grave!'

The boss felt like he had received an electric shock. Had this *sheedi* actually been to inspect the site before now? What was he driving at? What could he mean?

The *pattewala* was lost in thought for a while. Sitting on that stool year after year, as far as he was concerned time had already lost its significance. Forget about the day; he'd spend all night right here if he had to.

He spoke up. 'Boss, I'm not looking for anything from the board, or from the government, or from the contractor.'

'From me either?'

'Come on, Boss, you're like my parents to me. If I need to, I'll ask. I'm your *put-vaangoon*! If a son can't ask his father, then who *can* he ask?'

The boss chuckled, 'I think maybe you've become a communist.'

'How do you figure?' the *pattewala* asked innocently.

'It sounds like you're speaking out for the cause of the people of Goth.'

'Me?'

'Yes.'

'*Me?*'

'Yes, you!'

'God forbid!' the *sheedi* said touching his earlobes and running his index fingers down the sides of his nose. 'I'm a Muslim, Boss, a good Muslim. How could I be a communist?'

Laughing a little too hard the boss said, 'Okay. Should we go?'

'Wait a minute, hold on. What we can talk about here can't be discussed in Goth or the office.'

'Then get on with it!'

'I will, just let these ox carts pass by a little.'

Three carts, with three dogs walking in their shadows, were slowly approaching. Suddenly, out of the bushes, another dog emerged and began barking at them. The three tame dogs held their peace and kept walking in the shadows.

After the road had cleared the *pattewala* said, 'Look Boss, even I know that the commision for a contract is eighteen percent. First it was nine, then twelve, then sixteen, now it's eighteen. What do I get for each deal? Nothing. Once in a while fifty rupees, maybe a hundred. These clerk-types make it seem like they're giving a handout to some beggar.'

The boss wrinkled his brow.

'Now twenty-five percent is the deal for this contract. What will I get? The same measly fifty or hundred rupees.'

The boss hardened his tone and said, 'You know you're talking about bribery.'

The *sheedi* repeated his ears-and-nose touching routine. 'God forbid! No Boss, I'm just asking for what's right. Come Judgment Day no bribe-taker is forgiven, and none gets to view the Prophet's face, either.

'All these people are needy, but *I'm* not. You've only got one wife and one mother, and look well you take care of them. But I've got an entire family to worry about—my wife, her mother, my father-in-law, five daughters, four small boys, my invalid father, my younger brother's widow, her four girls, five small boys....'

'Okay, so?'

'So nothing, just listen right now. You make more than I do, but even still, do you know all the things that you get for free, but I don't? I can't even get the bill for my medicine approved. Okay, you're an honourable man, I'm not talking about you specifically. I'm giving an example. When some supervisor's medicine bill gets approved, he can even take a cut from *that* to put aside for a rainy day. But me? I sit in the same place for thirty years, like a dog. I wag my tail and stir a little dust over here, then I go sit down over there.'

'So what's the solution?'

'Boss, you've finally gotten to it. Look, fifty or a hundred a month won't do me any good. Before you depart this world God will provide you with a high position. Just do this much for me: let my cut be figured the same way everyone else's commission is. Even if that's only a single paisa on the rupee. Otherwise...'

'Otherwise?'

'Boss, even a little ant will bite, if he's being stepped on.'

As the sun climbed past its zenith, the sunshine was now falling directly on the windshield, and it had become hot. 'A fine place to just sit!' the boss thought to himself. Even a bridge on the canal would have been better. This was no place to stop. But then it occurred to him, 'Did I stop the jeep here, or did the *sheedi* make me? Maybe it was me.' The *pattewala* broke his reverie.

'If a man knows what his monthly earnings are supposed to be, he can keep his wife and kids happy. He can plan ahead—how much to spend, how much to save.'

The engineer knew just what the *pattewala* was talking about. Didn't he write the exact same thing to his wife's brothers all the time? What were they thinking, sending off a few thousand whenever the fancy struck them? How could anyone stick to a budget or make any plans? They should have been careful to calculate their sister's share after each harvest and send her exactly that much. But who listened?

The *pattewala* spoke aggressively to the boss, like an animal shaking its prey. 'So tell me, Boss. What's your decision?'

The boss was still angry at his brothers-in-law. He said hotly, 'Get out of the jeep! Do I work for you? I'll call you in to explain yourself when I get back to the office.'

'That's up to you, Boss,' the *pattewala* said with complete calm. 'But let's go to the site first. The whole population of Goth is there, and they're not very happy with the government.'

'Why is that?' the boss said, calming down a bit.

'The people say the water pump, the one we drilled ...'

The boss felt like he had been stung by a scorpion. 'What water pump?'

'The one some outside agency—maybe that children's fund—gave as a gift. They know the water has to be taken from the canal to fill the tank. But as it is, when they go there, there's already no water for their livestock....'

The boss gestured for the *pattewala* to be quiet. Trying to calm himself he said, 'How many did you say were in your house?'

The *pattewala* began to count: 'Two, three, five, and five more, ten ...'

'Absolute sonofabitch!' the boss said to himself, and peeling off a few of the largest bills he had in his pocket he gave them to the *pattewala*. Mimicking his speech he said, 'There, now, that should do. Like everyone else, you'll get a steady cut. You'll be able to plan for your future.'

He started the jeep. The *pattewala* pulled the last two cigarettes from his pack. He crumpled it up and took aim at a lizard stopped in the middle of the road, craning its head to look at something—maybe the traffic, maybe a companion it left behind.

As they approached the village, the engineer saw some women with large clay jars on their heads coming up the path that led from the canal. Waking from their naps under the trees, some dogs came running alongside the jeep barking their welcome.

The *pattewala* said, 'Boss, by your graces, I know a lot of people around here. If they say to spend the night, don't refuse. They have all the amenities—rooms, everything. And then ...,' he smiled. 'You're still a young man. And your wife's not home. You can get anything you want here. And we're not far from the border. The best of everything—European, whatever you want. You're like my parents to me. You just say the word. I'm your *put-vaangoon!* Don't be shy, *munjhna saa'iin!*'

—*Translated by G.A. Chaussée*

Kanha Devi and Her Family

One of the houses in Memon Para belongs to Kanha Devi. The residents of the neighbouring houses are also Hindu. This Hindu enclave, created by the intersecting geometry of lanes and bylanes, looks like a tiny island surrounded by a sea of Muslim dwellings. These people, the majority of them at any rate, have been living here for as long as anyone can remember. Once in a while, though, a relative from Bharat washes up on the tiny island like a wave, and then recedes. Life resumes its old rhythms, both inside Kanha Devi's household and out.

In the narrow bazaar, shy of daylight, wooden handicrafts are skillfully carved and sold. They also do some brasswork here, and the bazaar is always well stocked with homespun cloth from Sindhi villages. When people jostle about in the crowded bazaar, rubbing shoulders, it is impossible to tell their communal identity. Just about everyone here uses *salaam,* the Muslim greeting, and the Hindu custom of touching the feet of one older than oneself is so widespread among the locals that one never knows which of the two—the one touching the feet or the one affectionately placing his hand on the other's head—might be Hindu and which might be Muslim. Both may be Hindu; then again, they may just as well both be Muslim.

Certain items, though, are never sold here: miniature idols used in worship, copies of the *Bhagavad Gita,* and those big, coloured pictures, exquisitely printed on fine art paper, that the Hindu inhabitants of the neighbourhood use to decorate their homes. Some of these, in heavy gilt frames, hang on the walls of Kanha Devi's sitting room, and some, smaller in size, on the walls of the small prayer room directly behind the sitting room. A statue of Shiva and a brass oil lamp adorn a niche in one of the walls of the prayer room.

One step into these two rooms and you feel transported into quite another world—of long ago, inhabited by Krishna Gopal, the sly butter thief; Ganesh, with his elephant trunk; Shiva and Parvati; and Duhshasana trying to disrobe Draupadi—but an unseen blue hand helping her, spinning Draupadi around, feeding out length after length of sari, like the hand of a kite-flyer in the heat of the contest, letting the string spin freely off the spool, causing his seemingly victorious opponent to give up the match in sheer exasperation...exactly what happened with Draupadi. 'Unless a woman is simply bent on a loose life—flitting around like a kite with a cut string—she always remains safe behind the folds of her sari, until such eager hands as those of Duhshasana are ready to admit defeat.'

And as she recounts all this, Kanha Devi herself becomes chastity incarnate.

And this picture here, it shows Bharata removing the wooden sandals from Ramchandraji's feet, so that he may install the sandals on the throne and rule by proxy until such time as Ramchandraji has returned. One cannot even imagine such a thing today. Why would the proxy relinquish the throne once he's got it? Better yet, why would anyone give up the throne and go into voluntary exile in some forest in the first place? And leaving his sandals behind with his brother, at that!

Kanha Devi's sister fasts every Monday in honour of Lord Shiva. And Kanha Devi's brother-in-law—Persumal, who is well only half the year and is seized the other half by terrible fits during which he rants and raves, shouts and screams—is in the habit of getting up before dawn and feeding the birds. At this time, the birds come in flocks and alight on the low wall around the roof of the house. Persumal feeds them and recites from the *Bhagavad Gita* somebody had brought for him from Bharat some eight or ten years ago.

This usually means that he is well these days, and the whole family can breathe a little. Which is not the case when he is in the throes of a fit, for then he begins to curse his *wanyas*—that is, his own caste people—up one side and down the other. It would seem that he recites the *Bhagavad Gita* when he is healthy; but it also seems that it is only during the bouts of anger and insanity that the sacred text actually begins to unravel its meaning to him.

Persumal calls his father and older brothers *wanyas*. These people lend money at terribly high rates, unabashedly eat meat, liberally consume alcohol, and smoke marijuana—in other words, they do just about everything the *Bhagavad Gita* considers reprehensible.

When Persumal is giving them a tongue lashing, Chandumal, Mirchumal, Dhumi, Sita, Ashok, even Kanha Devi herself— everyone feels terribly embarrassed. They touch the tips of their noses and then pinch their earlobes in repentance and say, '*Ishwar* has given him so much knowledge, if only He had also put a brake on his brains!'

'By giving Persumal so much knowledge Allah has really made our life miserable.'

Now, who would say these are Kanha Devi's words? But they are.

Just as these people freely use the word *salaam* in their conversation, so they also use such other Muslim expressions as *Allah* and *insha'allah*. And nobody stops them. By having a different faith one doesn't also come to have a different creator!

Usually on such occasions, when Persumal is in a temper, the Muslim neighbours take him over to their homes and do everything to calm him down. One takes out a *ta'wiz,* a charm, and puts it on him, another wraps a black string around his thumbs and wrists, and some woman brings holy water from the sanctuary of one of the local saints and tries, by earnest entreaty, to make Persumal drink it. Persumal, otherwise a strict Vaishnava in dietary matters, finally drinks from her glass, and does not insist, 'No, I shall not drink from your glass because you people are meat-eaters.' Thus in madness, he manages to tell only what is truly blameworthy in man, never that which man has fabricated in order to put some above some others.

Their Bharati relatives, who visit them from time to time, have remained in the last thirty-odd years since Independence exactly where Manu had left them centuries ago: bound by caste differences and distinctions. The children of many such relatives now read and write only Hindi. During visits when these children play with their Pakistani relatives' Muslim neighbours, in Kanha Devi's inner courtyard no less, the older women from Bharat, unbending in religious matters, just gawk at this outrage with unbelieving eyes, though Kanha Devi herself finds nothing wrong with it. Children aside, even the adults in Kanha Devi's neighbourhood seem to have in one fell swoop scaled the restraining walls of untouchability and caste differences, and stepped into a refreshing open space where the son of the Brahmin Sri Ram is about to marry the daughter of the *wanya* Dhumi. Nobody loses any sleep over this glaring infraction of caste rules.

And not just that. Even the Shudras of this place freely participate in the Ram Lila festival. Every one of the last six or seven years, Okha's son has been playing the rôle of Raja Ramchandra in the Ram Lila play. Stranger still, at college, this fellow and the Brahmin and Muslim boys often drink unreservedly from the same glass, no different from the children playing in Kanha Devi's courtyard. In spite of her deference to Hindu customs, the thought never crosses Kanha Devi's mind to reserve separate glasses for the Muslim children, nor to serve her orthodox Hindu relatives in separate dishes.

So when Kanha Devi got her son married to a Bharati woman and brought the bride over to Pakistan, a strange tension swept across her home, although she remained unaware of it for quite some time. She had spent all her life rolling out *papars*, a daily staple in her family fare, doled out to the neighbours and served up as handy snacks to guests. Likewise, she would spend the whole year preparing *achaar*, and on festival days make pancakes and other sweet dishes. Preparing two kinds of dishes was her responsibility: one a sort of poultry, seafood, and red meat combination, the other strictly vegetarian. In preparing the latter, she would be careful not to use the same ladle with which she had earlier stirred the meat gravy. It seems these chores had been her charge from as far back as one cared to remember, from when she lived at her father's, just as later at her husband's.

Persu's wife was none other than Kanha Devi's own younger sister. Living with a crazy husband, she had become half-mad herself. Burn incense sticks or fast in honour of Lord Shiva, that's about all she knew. But nothing worked: she remained childless.

In that crazy environment, where the males in the family indulged in gambling throughout the rainy season and waited eagerly for the festivals of Holi and Divali to get drunk, Kanha Devi alone knew what it took to run the house.

Kanha Devi had succumbed to the fatigue of old age by the time that Damayanti, her daughter-in-law, had arrived. She had it all neatly figured out: 'The minute Damayanti sets foot in the house, I'll leave everything in her care and retire. All day I'll sit in my swing-seat and chat with the neighbourhood women, chew a *paan* maybe, or smoke a cigarette, and without so much as ever even lifting a finger be grandly served by her at meal times.'

Damayanti took over all right, but what she did not do is tell the old lady: 'Okay, I'll do the work, all of it, but you must promise to guide me. You must tell me exactly how many *papars* to roll, how much to spend and when, and what dishes to cook on what day.'

Only much later did Kanha Devi realize that Damayanti had already been stewing in her own juices for two, maybe three months. It had begun to annoy her that the neighbourhood children freely cavorted around in the courtyard. And yet if you sounded her out, you would know that Damayanti didn't really dislike children, though she did take umbrage with their barging in and out of the kitchen whenever they felt like it.

On the other hand, Damayanti didn't dislike visitors from Bharat. She would receive them with open arms, never mind the fact that they were driven to this country not because of any great love for their relatives, but by the desire to visit the sanctuary of their patron saint and offer up their votive dues, exactly as Kanha Devi herself had once gone to Ajmer in Bharat.

And visitors from Bharat brought gifts: colourful *bindiyas* and saris (which Damayanti was so accustomed to wearing herself, but which the other girls at her in-laws' did not wear), film magazines, and lots of gossip from over there. That of course was understandable. But Kanha Devi had no idea that Damayanti should find the old acquaintances of the family so exasperating. Damayanti just couldn't bring herself to drink tea from the same cup used earlier by a Muslim guest, or, for that matter, a Hindu guest ignorant of the importance of caste differences. Then again, names such as Ramu, Shyamu, and Gopi were too fuzzy to indicate one's true caste origin. Worse still, just about anybody walked in uninvited, expecting hospitality.

Now not only did Kanha Devi herself visit her Muslim friends in their homes but also dragged an unwilling Damayanti along.

It seemed that true *dharma*—or at least as much of it as Damayanti could understand by her own lights—had all but vanished among the local Hindus, and the pictures of Sarasvati, Lakshmi, Krishna and the Gopis had been hung on the wall merely to invoke their blessing and protection. Absence of hatred for those outside the fold was as good a sign as any, Damayanti thought, that the local Hindus had gone slack in their faith.

Then again, she could lick any one of them hands down when it came to true religious knowledge. Didn't she know the *Bhagavata* by heart? And the correct meanings of Hindu names? It would seem that the myriad communal riots and clashes of high-caste Hindus with the Harijans, the untouchables, in her native Bharat, had quickened that religious nerve in her that feeds on hatred.

How many local Hindus had gone to Bharat on pious visits and pilgrimages? How many?

One night she asked Kishan Chand, her husband, 'Are we going to rot here forever?'

Damayanti's question exploded like a bomb. Kishan Chand, who had just dozed off, was startled, and he asked, 'Forever— what do you mean?'

'You *like* it here?'

'You don't?'

But Damayanti remained silent. Kishan Chand, sensing the palpable tension around him, tried to clear the air with a lighthearted joke. 'Don't you like me? Ah, I get it, you've got somebody over there.'

'I'm not talking about you. I'm talking about *here*.'

'Here is ... here,' Kishan said. 'I'm from here. If I don't like it here, who will?'

Damayanti was sitting with her head tucked between her knees. She was uttering every word with the greatest circumspection, hesitantly, just as every girl does when she talks with her husband about some worldly matter for the first time, a matter which invariably ends up being the opinion of her in-laws.

'The fact is, the customs here are a little strange,' she said.

'And yours were different there?' Kishan asked.

'Yes. There we mixed only with our own kind, our equals. And nobody dared barge into the kitchen with their shoes on, as they do here. Nor did we cook meat.'

'But whenever I visited your folks, I always got to eat meat,' Kishan said.

'Not in our house. Maybe in other people's houses.'

'So where do the men in your family go to eat meat?'

'In restaurants,' Damayanti laughed.

'I get it. You'd rather we ate meat out in restaurants here too, is that it?'

Time and again Damayanti tried to get on with the subject, but it seemed as though she and Kishan were talking on two different wavelengths.

Finally, Kishan Chand said, 'You know what I think? If you had been born here, then you too would be like mother: you'd obey your religion and not hate others for obeying theirs. Anyway, why would I want to abandon this country? This is a land of *opportunity*!'

'What's that?' Damayanti asked naïvely.

'Let's just say that I'm in no mood to emigrate to Bharat. I'm happy here. I've grown up among these people and consider them my own. Your misfortune is that you grew up in an environment full of instigators, people who keep themselves in business by stirring up members of one faith against members of another, and send one caste against the throat of another.

Lucky for the politicians! Even in this day and age, they can find enough ignorant people to shore up communal unrest.'

Many times thereafter, when alone with the younger members of the household or with her husband, Damayanti would take exception to their use of the phrase *insha'allah*. She even set her cup apart. When forced to accompany Kanha Devi to a Muslim friend's house where they would be offered tea sent for from the neighbourhood restaurant, Damayanti would find some pretext or other to leave without drinking any.

One morning Kanha Devi's husband, a cotton merchant and moneylender, returned home unexpectedly early. When asked about it, he said that communal riots had erupted in some industrial town in Bharat.

Kanha Devi, as was her wont, said disinterestedly, 'Well, then, shall I start packing?'

In the kitchen Damayanti's hands suddenly stopped what they were doing. She didn't catch the note of sarcasm in Kanha Devi's voice.

And it had gone on this way in this household for the past thirty-odd years. For although Chandarmal did do business here, he always looked like a bird poised to take wing any minute. He thought it unwise to tie up his money. But the other *wanyas* carried on their business undisturbed: one ran a bakery, another a restaurant, another made movies, and yet another worked as a contractor. Chandarmal alone looked jittery, always in a big rush, as though he would miss the train. Whenever news arrived of a fresh communal riot in Bharat, he would right away hop on his swing-seat and start pulling nervously at the hair on his chin. People say that he too had in him a streak of the same illness which afflicted Persumal.

The same nervous tension gripped him that day. When he arrived at the bazaar, he found the others busy at their work. Alone Lala Ram, the photographer, asked him in a hushed voice, 'Did you hear the BBC this morning?'

'No, why?' Chandarmal asked, all keyed up. 'Did you?'

'Communal riots ... on a large scale.'

'Where?' Chandarmal asked, although he knew the answer.

'Bharat, *chacha.* Where else?'
Each looked into the other's eyes.

A little later Chandarmal's ears began to buzz with the noise of the paper boy shouting the headlines: THREE HUNDRED MUSLIMS SLAUGHTERED IN HINDU-MUSLIM RIOTS! POLICE OPEN FIRE ON MUSLIMS! Hidden behind these words were the sowers of dissension and chaos, those who turned communal riots into a roaring business. Hawkers were happy that they would be done early today. Chandarmal alone had no idea quite what to do. He walked to one end of the bazaar and then back to the other, hoping to gauge the people's mood. But people were preoccupied with their own worries: one had to take his polio-stricken child to the hospital, another had a court hearing to attend. Somehow the Hindu shopkeepers appeared more preoccupied with their work today than usual, deftly avoiding the eyes of others.

Deepak, the tailor master, was marking the material spread out before him with a piece of blue chalk, his head hung low, while his son busily took the measurements of a Baluchi youth.

Damayanti felt the night growing oppressively long. Kishan made no mention of the communal riots. He had gone to a movie with some friends, and when he got back, he went straight to bed.

In Kanha Devi's small two-story house, crammed with some fifteen people, Damayanti was feeling herself perilously alone, expecting something terrible to happen any minute. Every whistle of the night watchman startled her.

At dawn when the sound of *azan* arose from the neighbourhood mosque, Damayanti felt that fate had brought her into a cul de sac from which she couldn't possibly hope to escape. She was the animal tied to a stake and beaten to death. How different was her present from her past, when she still lived back in her own country! Over there she wouldn't have given two hoots about the communal riots, she wouldn't have lost any sleep at all over them.

All minorities, like an orphaned child, fear the worst, even in their dreams.

When Kishan stirred in his sleep and said, 'What, up already?', Damayanti quickly answered, 'I never really slept.'

Kishan, still in bed, threw his arms around a swollen-faced groggy-eyed Damayanti, who was sitting up in bed beside him, and asked, 'What's the matter? Don't tell me somebody drank from your glass again.'

'That happens every day. How much can one avoid ...'

'Then *don't*!' Kishan said, lifting a lock of her hair, and then added, 'Join the others. Mix with them.'

Damayanti freed her neck from his coiling arms and said, 'Come to Bharat with me. I will never be able to sleep peacefully in this country.'

'Why? Do beds have thorns here?'

'This isn't even your country. It's *theirs*.'

'Theirs—who?'

'Those who surround us. Who created this country in the name of religion.'

For the next few minutes Kishan strained as though trying to read some invisible writing on Damayanti's face. He said, 'Look at it this way: if a woman can be wife to one and mother to another at the same time, then why can't the same piece of land be held dear by some, because it was gotten in the name of religion, and be respected as a motherland by others? Tell me, when you become the mother of our child, will you stop being my wife? Or must we have two Damayantis?... Only then would it make sense to think of one as mother and the other as wife.'

Damayanti laughed. But her worried heart kept pounding in her chest. Never before had she heard the sound of *azan* come from so nearby.

Early in the morning when her chores brought Damayanti to the rooftop, she found Persumal feeding the birds as usual. This was something he never failed to do, not even when sick.

Everybody is nuts in this family, Damayanti thought, whether it was her mother-in-law, her husband, or his uncle Persumal. Why else would anyone worry about feeding the birds at this outrageous hour?

Instead of finishing her work and returning downstairs, Damayanti decided to stay a while and watch Persumal, who had meanwhile joined his hands to pay respects to the rising sun and was mumbling some pious words.

When Persumal was done praying, he asked her affectionately, 'You want to ask me something?'

'As a matter of fact, I do,' Damayanti said. 'Yesterday there was a riot between Hindus and Muslims ...'

'Where?' Persumal asked without much enthusiasm.

'In India.'

'So, what else is new?' Persumal uttered these words as though the occurrence was about as important and frequent as a common cold.

'Don't you ever think of moving over to India?' Damayanti asked.

'Your father-in-law—my brother—does. I don't.'

'Doesn't it scare you? What if somebody provoked these people? We're surrounded by them, you know. I'm so scared I stay awake the whole night long.'

Persu lifted his hand and pointed at something in the sky, and asked, 'What is that?'

Damayanti looked into the space and replied, 'Why, a minaret, of course.'

Persu shook his head and gestured toward the pigeons pecking at the grains in front of him. 'You know something?' he said. 'These pigeons roost in that minaret at night.... More than this I shall not say; in fact, I'm not permitted to say.'

Damayanti wanted to ask who had prevented him, but decided to keep quiet as Persumal had already become joyfully absorbed in his worship.

Downstairs Damayanti told Kanha Devi about it, who, instead of laughing at the matter, said with a feeling of profound respect, 'Let people call him what they will. But Persu is no crazy. I say you can go through the whole city, let alone this neighbourhood, and never find a soul more enlightened than he.'

—*Translated by Muhammad Umar Memon*

A Man's Country

Mobinurrohman was one of those people who are really quite decent but are somehow spurned by those around them.

He would often be seen at work on the ship's bridge, or sitting alone on the closed lid of the hold twining rope. No one had ever heard him hum a note or utter a cry, nor had his breath ever smelled of alcohol. Generally, too, sailors from his part of the world drank very little or did not drink at all.

This, we thought, had to do with their miserliness, neither themselves drinking, nor ever buying anybody else a drink, nor even rushing from the port to the town where women from all over the world sat waiting for us, the sailors.

In short, Mobinurrohman, who his co-workers called just Mobin, was not one of those sailors of whom it is said that they have a dame in every port.

We were a fairly free and easygoing bunch, swilling drinks and always drinking in company. Our Purser would gladly get a tab going for everyone's purchases; even then, we were always willing to lend a helping hand to any of our mates who didn't have cash on him, who would then return the favour, later on, if the need arose. Sometimes it would so happen that you would be sitting in a bar drinking with a lady friend, and putting your hand in your pocket suddenly discover that there wasn't a copper in it. At such times, if any of our shipmates was about, he would be called over. He would realize the seriousness of the situation, come up with enough cash for the last drink and say, 'Hurry up, I'm waiting for you outside.' You would be wise to do as you were told—by one who was still in control of his senses. Otherwise, after his departure, neither would the bar-boy look at you, nor would your lady friend keep you company for long. And, also, the whole business of getting to the ship would still be there—the taxi fare, the walk to the port, the entry into the unsteady sloop on staggering legs, the walk to your cabin on the gangway of the rolling ship. Often, for hours you didn't know who you were or how you came to be where you were.

But these sailors from East Pakistan! I don't recall if any one of them was ever helped to the ship, even once. We used to joke about them: every one of them drinks a bucketful of water before going ashore—as children do at dawn before the fasts of Ramzan—lest he have to buy a lemonade in town. We believed these people also used to eat a full meal so that they wouldn't feel hungry during the few hours they had to spend in port.

Before the ship docked we always arranged to hide some liquor, which we recovered as soon as the customs officer had left the ship. We would drink what we needed and take the rest out to sell. Some of it had to be given to the duty guard at the gate: sometimes a packet of cigarettes would do, or a bottle of beer or stout which would cost only a few pennies on board. Sometimes some of us who were off duty, knowing that in the next port of call liquor would cost less than a glass of water, would go on the wagon a day before the ship came into port. There you would be able to get the best rum and the most potent brandy at the cheapest prices, one sip of which would be enough to transform the ugliest hag into a delicate beauty. When we came back from such ports we would all be drunk, each one swinging in his hands a bottle or two of rum or whisky.

But whenever the ship docked in such ports where there was nothing except drink and women, Mobinur's mates would either be found sleeping somewhere on the deck, or, if they were off duty, trying to catch the fish that would be circling the ship if the cook happened to have dumped leftovers into the sea. They were really good at that.

Passing by him at such times I would say, 'Mobinur, what are you trying to catch? There's no *hilsa* here.'

He would smile, and if he were in a good mood, answer, 'But *hilsa*'s "brather's" here. I "vaiting" for them.' His being in a good mood meant that no letter from home bearing any bad news had arrived, that his mother was well, his father too, also his sister and three younger brothers, as well as his wife and kids; even his older sister and her husband and children, who lived in some small town with a name like Munshi Ganj, or Narain Ganj, or Gopal Ganj, at a distance of about forty or fifty miles, three or four rivers away from Mobinur's own hometown.

His brother-in-law was also a sailor like him, but on another shipping line. I had never met him, neither in those days when Mobinur and I worked together, nor later. But I would often see Mobinur's face suddenly light up when he recognized the stacks of a ship going far in the sea and exclaim, 'That "seep" is of my "brather's" line!'

Once, our ship came—in the seamen's language—alongside a dock in a port. Two or three ships ahead of us, a ship belonging to another line had already docked. Seeing that ship Mobinur went simultaneously into a fit of joy and a flurry of activity— perhaps because it was a good port and that meant that there were going to be ore cranes, more on-shore workers, quicker work and higher per-hour charges for the ship's stay in port. At such ports, ships would dock only for a couple of hours, and very few of us could get leave, and then only for a very short while.

Mobinur's joy, in my view, had much to do with his stinginess. The ship had come alongside; that meant that now he would not have to pay the sloop oarsman to carry him to and from the shore. This he would have had to do if the ship had anchored farther out to sea. But it was no big deal. If he really needed to see his brother-in-law or find out about him, what was a few shillings? People from my side of the world never bothered about such small expenses.

Mobinur was the most niggardly person I knew. The European Second Officer who was looking after the off-loading of the ship's hold told him again and again, in an effort to get rid of him, 'No, no, Mobin. You can't leave. Go work. Duty. Duty. Go work,' while he was also ordering the crane operator to lower or raise the cargo.

Expending his entire stock of English, Mobinur tried to make the Second Officer understand that he would be back as quickly as possible. His sister's 'hosband' might be there on that 'seep,' or there might be some letter from home. Many people from his part of the world were there in the 'corew (crew) of that seep.' There may be someone there who might even know his brother personally.

Because of the heat, the Second Officer had undone all the buttons of his shirt and was downing can after can of cold German beer. He was unable to understand why these people— he meant all of us—were so anxious about letters from home, to find out how everyone was doing back there. If the Europeans were like that, they would be unable to work in peace for half an hour, for one had one sister in Australia and another in Canada; one's mother was in Holland and one's father in the Congo. We learned such things from the Chief Stewards on different ships. A Chief Steward was usually a man from our own part of the world, a second class officer on the ship, who could speak English as well as some of our local languages. Chief Stewards usually had a share (as did some of the European officers) in the minor swindles that we carried on at sea, as when we smuggled small amounts of gold here and there—not shiploads of it; we did not do things like that.

Gently the Chief Steward tried to make the Second Officer understand that the ship's Captain and the Chief Engineer had already gone into port. But before he could finish his sentence, the Second Officer blurted out: 'Oh, what else can they do in the port besides...,' and leaving his sentence incomplete, he started taking swigs of beer from his can, all the while ordering with the movement of his fingers the crane operator to move the arm up or down.

Then, when the Chief Steward tried to recommend Mobinur's leave, the Second Officer swore at himself and said to Mobinur, 'My sister also home. She also not all right. I no die. Understand?'

Mobinur's dark cheeks reddened instantly and he asked the Second Officer if he had used a swear-word for Mobinur.

'No, no, not for you,' the Second Officer cooled down suddenly and putting his hand on his heart said, 'I swore at myself. See the work going on and me, and me...'

The Chief Steward explained to Mobinur, 'Don't mind him. He's drunk. Has gone half crazy handling the crane. He's swearing at himself, as sometimes even we do when there's too much work, like 'bugger me' or something.'

Looking repeatedly, one after the other, at the faces of the Second Officer and the Chief Steward, Mobinur said, 'There in my "contree" cyclone, this monsoon "seajen"; men die. Thousands. Ten thousands. "Sildren". "Booman". Old men. No home. "Saar", there all young men go work. "Feesing". On "seep". "Bery" far. On "reever". Who make house? I "burry" "bery" much.'

The Second Officer swore again, 'Akh, Mobinur, your country always men dying. They no want to live. Always cyclone. Always famine. Flood. Damn it.'

The Chief Steward wisely refrained from translating this comment, or there would have been a scuffle between the Second Officer and Mobinur right then and there. And since this was a matter happening on land, not at sea, some port official would have to settle the issue, not the ship's Captain. At sea, the Second Officer could have consulted the maritime laws and punished Mobinur as he thought fit; he could have spoiled his report, noted in his passport that he cannot control his temper and doesn't obey commands, but here at port, he was helpless, while Mobinur was ready to wage a war. Actually, at heart, this Second Officer was not a bad fellow. Whenever any crew member fell ill, he would go to inquire after him ten times a day. And if his case was serious, he would arrange to have him admitted into a hospital in the next port of call, and thus prove himself to be more loyal to the crew than to the company that had hired him. More often, the shipping companies, whether native or European, tended to be worse blood suckers than even the usurers.

The goods were now being hauled out of the lowest hatch. For some reason the crane operator stopped working. Far away, on a narrow gangway, some stevedores were prodding and shoving cattle onto the ship.

The Second Officer threw the empty beer can into the sea, lit a cigarette and said, 'Akh, as far as I care, he can go to his country, if he likes, and save it from cyclones. What is he doing here? What is his government doing?'

'What Sahib saying?' Mobinur said apprehensively. 'Government issue new orders?'

'Nothing, nothing,' the Chief Steward said, 'He says go find out if there's been any cyclone.'

'Thank you, "saar",' Mobinur answered, quiet and subdued. He wasn't convinced by the Chief Steward's translation.

'Is there cyclone again there?'

'Go, go,' the Second Officer pushed Mobinur by the shoulder. And then asked the Chief Steward to tell Mobinur in his own language to go have a few drinks. It was because he didn't drink that he was always worried about cyclones, about ferries sinking and about people drowning.

That night I couldn't see Mobinur. The next morning when we came out to swab the deck, that other ship and port were miles behind us. Now there was only the ocean, or the river, as we used to call it, all around us. The river, on this side, was always stormy. You didn't feel like eating, drinking, smoking or even talking to anybody. We wore full rubber boots and were scrubbing the deck with mops. Mobinur was throwing sea water on the deck with a hose. I greeted him with the Bengali word *bhaalo*. I would use this word more as a 'hello' than to find out how he had been feeling. He just nodded in response.

After lunch I took two aspirins from the Second Officer and went to my cabin to rest. My head was still heavy. The Second Officer said when he gave me the tablets: 'Remember, Abdul, you are a Mozlem. You are not supposed to drink, and you drink too much.'

I answered, 'Sir, you also remember: like Christians, Moslems are also of two kinds: good Moslems ...'

'...And bad Mozlems,' he completed my sentence. On this particular voyage shipmates from my part of the world were pretty happy. Sher Afzal had discreetly delivered the pair of binoculars to the store keeper whom he had promised them and had earned about five pounds. Jeera had done the same with a camera.

The Third Engineer Tripstra, the Chief Steward and Fazla had managed to sell some gold. Fazla had gone to purchase it in the last duty free port and had handled the sale so circumspectly that no news of it had reached the next port. Tripstra was a white man. He had kept himself detached from the whole deal. If anyone had been caught, it would have been Fazla or the Chief Steward. Tripstra would have flatly denied having anything to do with it. But everything worked smoothly. The rest of us had either sold cigarettes or bottles of scotch. I, too, hadn't fared badly. But most of us had already squandered on women whatever we had earned. As the saying went: 'Seaman, funny life / New port, new wife.' Ever since leaving the port, we the 'fun-lovers' had gone around teasing those who had come from Mobinur's part of the world. 'Just as well, man, you didn't go into that port. Firstly, you wouldn't have been able to see the girls because you would barely reach their shoulders; secondly, the girls there are scared to death of dark skin.'

If somebody felt like carrying the teasing further he would say, 'The cows in your part of the world look as big as she-goats, she-goats as big as she-dogs, and you—do you know how big you look?'

Sometimes this sort of verbal sparring would lead to physical fights, because some sailors from Mobinur's world could also handle themselves well in such situations. One would, for example, ask a dark-coloured sailor from our side, 'Are you sure your father didn't come from our "contree"?'

In the afternoon I came out of my cabin. The pain in my head was gone; only the throbbing remained. I got a cup of tea from the storekeeper and sat on the closed lid of the hold in the shade. It was quiet all around in the ship. Only the sound of the sea or the splashing of waves could be heard. The workhands were either in the engine room or on the bridge; the rest might be asleep in their cabins. At such times nobody even listened to the radio which every one of us had. We had each bought one to take home with us. Some of us had three or four watches each, transistorized tape recorders, broad cloth for use in weddings, and ugly-looking rough-hewn gold rings—a common method of transporting gold home.

Mobinur himself and his mates used to buy German lanterns, stoves, Chinese umbrellas, and similar other objects of daily use with the little spending money we got on the ship. Also, whenever they got the chance, if they could buy the rupees cheap in any port, they'd sell their shillings and send money home. They did the same when they signed off from the ship. Every step of the way, they would like to take as much money home as possible. None of us could stomach their obsession with money—perhaps because none of us was himself capable of saving any.

A certain amount of money from everybody's salary was deducted at our Karachi office to be sent to our families. The shipping office was convinced that if we received all our pay on ship we'd squander it on wine and women, and our families would starve to death. But we knew fully well how much they cared for the welfare of our families and households. The company paid the shipping office our salaries in pounds sterling and dollars, which everybody knows are the most powerful currencies in the world. How were our families paid? In rupees. And guess who made the killing in all this? The one who didn't have to do anything. The capital that the shipping office invested consisted of men—of our labour; the gain they made was in dollars and pounds. And we were the safest capital: we couldn't run away anywhere. As the saying goes, one who has tasted of sea life once, returns to taste it again and again.

But we, too, were bastards of the first order. We wouldn't allow ourselves to be passed over that easily. Those among us who used to drink and had befriended the purser would spend so much by the time they reached the home port that nothing much would be left either for them or for their families. We would be told that no money has been forwarded to our families for many months. At that news we would feign surprise. That was sheer perversity on our part.

The shipping office was our lord and master. A sailor fears only two things: not being assigned to a ship and getting his report ruined. The condition of the sea didn't affect us, but being assigned or not to a ship was completely in the hand of the shipping Master—in other words, the shipping office.

The shipping office is a world by itself. Going in you seem to have walked into a carnival. You meet dozens of your old shipmates, some whose names you may even have forgotten. This one was with you on that ship; that one on another. You travelled to South America with this one, to Portugal with that one. This one comes from your own country; that one doesn't. Someone would be busy signing off, another joining the queue to sign on to another ship, and so on.

At the time of signing off in the shipping office, we would dutifully hand over whatever gifts we had brought—a carton of cigarettes, a bottle of scotch, or something small like a perfume bottle—to whomever we had brought it for, promising them we'd do better next time. We did that so that they wouldn't make it hard for us to sign on to a new ship after the break.

Sailors like Mobinur, on the other hand, always came for the signing off cringing and quailing. At the shipping office just about everybody usually looked better-fed than them—not exactly like a towering American movie star, but not nearly as stunted as these folks, either. And sailors like Mobinur were really the ones who had spoiled the people at the office. If someone, at the time of assigning them to the ship, had asked for a tea set from Germany, they would bend over backwards to obtain one, never considering the possibility of a substitute. They would be half dead with fear that if the Sahib got angry they'd have to stay for weeks in Karachi, because after signing off, these people had to go on their own expense to Chittagong or Navakhali. The company looked after their transportation only to the home port. In the Company's eyes their home was Karachi, but their actual home was elsewhere. They had to travel as passengers on another ship through Colombo, Chittagong, and God knows where else, and then on ferries and boats to get to their homes. They arrived in pretty ragged shape.

Our real peeve was that while we rested at home during the two or three months of the break, these people would be ready to sign themselves on another ship as soon as they had signed off from one, as if money were all that mattered and as if they didn't know that it could be spent. We used to smoke Craven A or 555; they used to smoke home grown tobacco that they brought with them, rolling their own cigarettes. But that's a different matter. Some sailors from our side, those who didn't have their homes in Karachi and who had to take the trains to get home up north, did not fare any better than those from East Pakistan. Not everyone's home was on the railway mainline; some had to take buses and hire tongas to get home. There were some consolations, however; there was no fear of drowning and when you reached home, you'd find it still there. No cyclone would have carried it off somewhere else.

I was sitting on the lid of the hold having tea when I happened to look at the other side of the lid which was in darkness. I saw someone sitting there. Could it be Mobinur, I thought, and called him in a dull, heavy voice. 'Mobin? Mobinur?'

He was probably knitting a cord-muzzle for holding cooking pots, or a coil-base for an earthen waterpot. He knew how to make many such things. He had once made something like that for me: it looked like a horse's tail and could be hung on the wall by a nail. In the hair that fell from its sides one could tuck in combs, brushes, etc.

Mobin kept quiet. I picked up my tea mug, and went and sat by him.

'Bhaalo?' I said.

His face was taut.

'What happened? Got a big "hort"?' I teased him. He had the habit that whenever he got a scrape or scratch lifting or lowering the cargo, he'd raise a caterwaul saying, 'I going die now! I got a big "hort"!' That was another reason, besides his stinginess, that people on ship kept themselves away from him—even his own countrymen. He did not talk much, but once he started it would be impossible to get him to shut up.

To avoid having to talk to me he said, 'News from my home not good.'

I should have kept my mug glued to my lips so as not to proceed with any further inquiries. The Second Officer was right: when was the news from his home ever good? It was either the cyclone or the sinking of the ferries. And these people were known the world over for their hunger. My own mother used to call me a starved Bengali if I ever hurried with my meals.

Then why did these twits not stay home to do the farming? Or make some arrangement to save themselves from the cyclones? At least they could build stronger houses which wouldn't fly away in high winds. Why did their boats always sink? Couldn't they make them bigger and stronger? Why didn't any meteorologist or someone who gave light signals from a tower stay in the village to warn people about the approaching cyclones, and tell them not to go out in their ferries? Everyone there was either a fisherman or a sailor; the rest had assailed our part of the world to look for work.

My own home is in Karachi, in an area which is not for rich people. But I know that in many areas of the city where well-to-do people reside, and in many other cities, there are thousands of people of Mobinur's nationality who work as cooks or drivers and are always cursing their fate.

But I couldn't bring my mug to my lips. Instead I put it down on the plank and asked Mobinur, 'Is it really bad?'

He nodded.

We sat there quietly. I was watching the sea. Big bloated waves like air-filled sails were forming and moving towards our ship.

Then Mobinur put the cord down, stood up and began saying his prayers right there on the plank. I felt as if he was not praying but crying. My mind was not working fully and the storm in the sea was affecting my stomach. I walked to the railing, emptied my mug into the water and, having nothing else to do, came back and sat down with Mobinur. A little later I felt as if he were talking to himself.

'What can one doing? It's God's "vill".'

'Did you say something to me?' I asked.

Then he couldn't seem to contain himself any longer, and I, whose senses had been dulled by the hangover, was startled by the tone of his voice.

'I said this God's "vill". What can a poor man doing? It is looking they "suit" (shoot) my sister's man. Many other men die there also. First cyclone kill some; others not yet build their house when people "suit" them.'

'Which people shoot them?' I asked startled.

'Your people.' And he glared at me as though I had been personally responsible for the deaths of his sister's husband and others, as though I were the one who had crossed the three seas to go kill them and was now hiding here after committing that heinous crime. By now he was again talking to himself.

'Now a "var" must happen. The man on German line "seep" tell me all. Thing being really bad. What can a poor man doing? Must fight to save life of "vife" and "sildren". Now nobody can stopping "var".'

My mind hadn't fully cleared yet. Also, I was not one of these sailors from our side who were glum and serious-looking, who never drank, who saved all their money and who constantly worried about the breakup of the country, talking about the treacheries of this or that race or the flaws in one or the other race. All this bothered me no end.

I knew that recently there had been a cyclone in Mobinur's part of the world; it was perhaps because of that that every Bengali had his ears tuned to the radio listening to programs, whether in Bengali, Urdu, Hindi, or English.

In actual fact, all the sailors of the world belong to one race, having the same language. Whatever radio station is on, every sailor hears what he wants to and can dance to the music of every country.

But I myself am one of those sailors who, once they are on the ship, completely forget at what price his family buys the flour or how many times a week or month they eat meat. I don't even notice the earthquakes at sea; what did I care for these little cyclones in East Pakistan. Even nowadays, if someone begins to discuss politics with me, I get uptight. In those days if someone asked me to listen to the radio to find out what was happening back home, I would answer that my radio catches only those stations where pop music is played, like Radio Ceylon. If it ever catches a news bulletin, it starts choking on it.

Now when I thought hard, with my heavy head, I recalled that the ship's crew for some time had been divided into two distinct groups: one that felt stricken about the fate of the country, and the other that was deeply anxious. Mobinur belonged to the latter group. He and his mates, it seemed to me, were constantly busy whispering to each other.

That was the first time I had seen tears in Mobinur's eyes. Perhaps among the many sailors from his side on the German ship there was one who knew Mobinur well. He must have exaggerated in telling Mobinur the story—that first there was the cyclone which blew away all the housetops; then a lot of damage was done by the government's people, who destroyed all the aid sent by the foreign countries. Who really knows what happens to the stuff sent from outside? Who hands it out? Anyway, it didn't seem to have reached the needy—the starving and the destitute.

'Why?' I asked him.

'So that all Bengali people starve to death.'

This argument was a little beyond me. Perhaps, as he had said, rice, lentils, blankets, and tarps had indeed been sent to East Pakistan by foreign countries, not by West Pakistan. Also, given the laziness and the insensitivity of the people in both East and West Pakistan, perhaps all the stuff had rotted in warehouses or at airports. But to claim that the government, or that the West Pakistanis, wanted all Bengalis to starve to death, or be destroyed by cyclones, didn't seem a very reasonable argument to me.

With a lump in his throat he said, 'I not knowing how my sister is now. How her "syeld" (child)? Everyone there afraid. Those who running to safety, "sot" with "masine-gun". My sister's man was being in boat. They fired and boat "ober-turn".

Those who swimming was "sot". What the man told me about was looking like my "brather", my sister's man....' Tears were running down his cheeks and neck.

I had to keep quiet for a while. Then I asked him, 'When was the last time you met your sister?'

'Three years ago.'

I was startled. 'You go home every year, and haven't gone to see your sister? Nor has she come to see you?'

He looked at me as one observes an ignorant child. Then wiping his tears he said, 'After signing off, when did I go Bengal last three years? You know how much it costing going there? A family can living three months in that much money.'

'And where do you spend your holidays then?'

'In Karachi. Doing small labour. What I can get.'

I got up quickly and without saying anything went to my cabin.

That was the day I discovered what the people of Mobinur's race thought of us. It was like what the blacks of South Africa thought of the whites who were few in number but controlled everything in their country—the government, the army, the navy, the air force. In fact, they considered us to be even worse than the whites in South Africa; they had the same view of us as the Uhurus of Kenya had of the English: they were sent from England to rule them, but after staying there their entire lives and after making money they had ravaged the place before leaving.

So what now? I thought.

Even I had quite a few relatives in East Pakistan, poor, unknown, unimportant people, those who at the time of the creation of Pakistan and India, because of being near East Pakistan, had gone to Chalna, Kishwar Ganj, and other such places and settled down there instead of coming to Karachi. If this idiot was right and a guerrilla war was going to start in East Pakistan, what would happen to those people? What England could they go to? Previously, whenever there was a strike in a factory or some other disturbance over there, some of them would run away and come to Karachi.

After this voyage, we had to undergo another and then sign off. In other words, we still had to spend another nine weeks on the ship. I don't remember having spent a worse time on the sea than that. People worked very quietly and, in their spare time, sat in their cabins and swore at the people of Mobinur's race. I think his people, among themselves, must have been cursing us.

In Karachi after we signed off, I saw Mobinur in the shipping office. An unusual thing was that he hadn't given the silverware drinking set to the person he had bought it for. He seemed a different man that day, not the Mobinur who used to scream, 'I got a big "hort"!'

His mates also had an air of insolence about them and were talking to everyone in the shipping office arrogantly, as one does to one's boss when one has made up one's mind to quit.

Passing him by I casually said *bhaalo* to him.

He said, 'I thinking you not say *bhaalo* to me anymore.'

'Why?' I asked.

'You "vill" find out. Let time coming.'

He and his mates were in a hurry to get to East Pakistan so I couldn't talk to him anymore.

But the time predicted by Mobinur really did come, and I did find out. Some of my distant relatives and acquaintances were killed in East Pakistan. The rest were trapped there in refugee camps. Whenever a letter from one of them managed to find its way to Karachi, having gone through various countries, it would still have Pakistani postage stamps on it, but now the stamps also bore the postmark of Bangladesh on them.

Those who were referred to there as the Dutch or the English had already moved to West Pakistan before the striking of the evil hour. The rest, impoverished and destitute, were slowly reaching Karachi one by one. Some had to spend the rest of their lives there.

West Pakistan had now become Pakistan, as if it had become higher in rank or station. Previously it was a part of a country; now it became the whole of it.

I quit my job at sea and was yoked to the family occupation that I hated—working on hand looms, sitting from dawn till dusk in the hollow of the weavers' loom and weaving cloth for loincloths. Time had also taught me to take interest in politics because many of our people who were trapped in Bangladesh had begun to escape through India and to trickle into Karachi. The tales of their hardships would regularly reach my ears: how much misery had been endured by which family, which ones were saved from Bengalis by Bengalis themselves, and how many were feeling unsafe even in Karachi, for the situation there was no better than in Bangladesh. Things seemed ready to fall apart, as if the end of the world were at hand. Hearing the names of places like Comilla, Barisal, and Nawab Ganj from their lips would remind me of many of my Bengali shipmates. Who knew how they were, whether they were even alive now or not!

Sometimes, when I ran into some Bengali in some part of the city, I would feel a sort of pleasure. He would be one of those who had decided to settle down in Karachi or in the Province of Sind. On my inquiring he would say, 'The condition there is really bad.'

It is no better here either, I would say to myself. Our chances were pretty equally balanced.

Bored with my work, one day without any particular reason I walked into the shipping office. This was many years after leaving ship. Actually, I had had a little tiff with my father and uncle. I was planning to do something different with my life. My uncle was also my father-in-law. It was a family squabble.

I ran into some old acquaintances there: Aslam, who had been a carpenter on a ship with me; Nazir, a boatswain; one or two people from the catering department; one *tindal*, and many sailors. Everyone seemed to have changed.

Going through the crowd, suddenly I spotted Mobinur. He saw me and lowered his gaze.

I knew that there were never enough home ports for these people to sign for work on ships. Some of them would even reach Calcutta without passports and visas and catch ships there. Obviously in Calcutta they would claim to be Indians.

I darted towards Mobinur, shook him by the shoulder and
said, 'Mobinur, is it you?'

He nodded.

This time the haughtiness of our last meeting was absent
from his face.

'How are your wife and children? Are they safe? And your
parents? And brothers and sisters?'

He kept nodding in answer to each of my questions. Then I
asked him, 'What brings you to Karachi?'

'Coming to signing on a "seep",' he finally responded.

'But you have your own Bangladesh now, don't you?' I
spluttered, as crudely and inappropriately as the Hindus and
Sikhs in India, I hear, used to say to Muslims, 'You have your
Pakistan now. Why don't you go there?'

Almost vengefully I asked him. 'You didn't go to
Bangladesh?'

He ignored the mercilessness of my tone and said, 'My
Bangladesh right here.'

'What?' I asked.

He repeated his answer, as if explaining something to an
ignoramus: 'My "contree"—this office, this work, right here.'

—Translated by Faruq Hassan

Together

'The poor thing. All alone in this frightful house. What will she do here after I'm gone?'

The old man just couldn't shake thoughts like this. One would take shape, only to have another come trampling in on top of it. Each new thought forced him to hollow and false self-consolation: 'This is all right, I should be doing this.' But the very next moment his first thought would lose its grip, and another would come to preside for a time in its place.

The old man and his daughter-in-law—just the two of them in the house. Everything he once had in his life was now gone; her life, however, had since childhood been devoid of anything whose loss would have caused her heartache. Her mother? She died when Raabi'a was small. The stories conflicted: one heard that it was really just a few hours after she was born, but also that it was only after she had begun to crawl.

In any event, everyone who might have been able to clarify things had themselves already passed away. She wasn't particularly interested in anything she heard about her father. It would have been different had it been about her brothers and sisters, but there weren't any of those in the first place.

What a frightening picture: to be born in a house with no doting grandparents, no solicitous uncles or aunts. More frightening still, to be born in a family with no one from her mother's side there to take a small girl in on the passing of her parents. But such was the generosity nature had shown to Raabi'a. So when she arrived at her in-laws' after her wedding, she had absolutely nothing—no expensive clothes or jewelry to adorn her body, no traditional gift from the groom, no dowry. Nor did she cry over leaving home; sad songs of departure are needed for tears like these, and the way these songs had been sung for her, well, it was no better than no singing at all. One night some girls had come and sat in the moonlight with the wooden drum that accompanies such songs, but they just tittered and giggled, as one does seeing someone accustomed to old, torn clothes, someone of low caste, dressed up in a suit for the first time. For those whose clothes are always nice, her immaculate outfit is no more than an occasion for some entertainment. On top of this, each song fell off after only half a verse, since each invariably made mention of the bride's brothers and sisters, her mother and father. Hearing shouts of 'You girls go to sleep!' coming from inside, they all got up silently and left.

After arriving at her in-laws' house, some of the distant relatives she had lived with wrote her a few letters, but only every month or two. When Raabi'a didn't respond, they too forgot about her. She acknowledged her indebtedness to them— they raised her, they provided her a fair amount of education— but all the same, she couldn't manage to keep this in mind. Time had shown her, like a kitten, through no fewer than seven homes, maybe more. When she arrived at her in-laws', she was simply one of those countless girls who rise at the crack of dawn, who wash up and prepare tea for the entire household, who, if they're in the habit, manage to perform their prayers as well. Otherwise it's just one chore after another until midday rolls around, then prayers again and the busy work for which the afternoon was especially created. After still more prayer it's evening, time to prepare dinner. Prayer again. Set the table. Serve the food, scour the dishes, then ask each of the adults if he or she might need anything before going to bed. After everything else, give the children their milk and finally lie down for the night.

People take a good daughter-in-law to be someone whose eyes fly open at the slightest sound. Others, who laze about late into the day, are said to be 'the essence of cow shit'. Just looking at them calls to mind the dung you see cows and buffalo leave behind in the street. Formless and flabby bodies, you can only *try* to wake them. Sleep-fogged, they indifferently start to nurse their crying children without first trying to find out what the real cause of the tears might be. When the child continues to cry even after getting a mouthful of milk, his mother gives him a 'pat' (which seems more like a slap); 'Go to sleep!' she says, and pressing his mouth even more forcefully to her breast, she begins snoring again.

Even after her wedding, Raabi'a remained exactly as she had been when she arrived in that house. She didn't grow fat, and there were no children to ruin her sleep at night. The chores she had done at her previous homes continued here as well—rising early, ablutions and prayer, the morning's tea, gathering the breakfast dishes, sewing and mending, lunch, dishes again, prayer. Then the afternoon's tea and prayer again, followed by a very long evening spent in the kitchen, during which, were it not for the refreshing break afforded by her ablutions and prayer, she would have felt herself to be some machine left running with no time to cool down, likely to burn up.

But after this break it was right back to the kitchen, and if anyone should come by to visit with her, the kitchen was her parlor. Fixed right there, she would occasionally draw her face from her work to answer some question she might be asked. Should her visitor feel a little closer to her, she might actually enter that world of heat and sweat called a kitchen and sit with Raabi'a. Fixed right there as well she would greet the woman who came round singing songs of God. This woman would sit with Raabi'a's mother-in-law and at her behest brightly sing her songs, one of which she sang at Raabi'a's insistence, too, at the tail end:

The bridge to Paradise is a difficult path
My insides tremble with fear
O Dear Prophet! Take my arm, support me!
I haven't fasted, I haven't prayed
My Lord will be angry at me!
O Dear Prophet! Take my arm, support me!

But the old lady showed up only infrequently; Raabi'a's daily routine continued without alteration—serving food as always, gathering dishes, prayer, and sleep that was usually dreamless, since she had neither a past nor a present worth mentioning. Occasionally she had dreamed that she was in a distant aunt's house, maybe on her mother's side, maybe on her father's.... She was neither happy nor unhappy; she just was. Often, when her husband was alive, it happened too that he would wake her up: 'Come on, get up. The day's well along.'

She'd sit up, rubbing her eyes. 'I was having a dream.'

'About what?'

'I was dreaming—I was dreaming that...Oh, I've lost it now.' And wondering at herself, 'What is it with me? I can't keep even a dream in my head for two seconds,' she'd gamely get out of bed to take on the new day.

Life provides the stuff we dream about, so when there is no stuff, how are dreams to be made? Raabi'a was unaware of this truth. After all, no one sees herself in a dream clapping out *rotis* as usual, or scrubbing dishes.

If she had ever made the holy pilgrimage to Mecca, perhaps she would have dreamed of herself performing prayers at the sacred sites there. Or perhaps she might have dreamed of those places she would have gone with her husband—a summer spent somewhere in the mountains, a trip to some important city. But even when her husband *had* suggested they visit such places, she shilly-shallied: 'Amma will be all alone.' As though she hadn't married a husband but rather an entire household, which she simply could not leave behind at will. All the fanfare of *other* peoples' weddings, the birth of children in *their* households....The sight of a baby boy, surrounded by relatives, free from all suspicion, full of trust...the barber deftly removing the foreskin from his 'pee-pee,' a few drops of blood left on the floor...the child's screams and sobs continuing to reverberate through the air...scattered bits of salty snacks and sweet rice that stick to the soles of your feet after parties...—All stuff insufficient for the making of dreams.

True, one time she did dream that her mother-in-law took her to visit a shrine. Despite her unwillingness, she was licking the dirt that the shrine attendant, telling her to open her hand, had put in her palm. But this dream was several years after her marriage, and she didn't think it proper to mention it to her husband.

The shrine attendant didn't really believe there was any chance that eating dirt would bring any sort of change into her life, and neither did Raabi'a. If anyone had any hope, it was her mother-in-law, who had put her through three or four operations, who had taken her into the presence of the pious and the Godly, who had given her cup after cup of medicinal herbal broth to drink. The performers of these kinds of small operations were themselves part of the world of shrines and seers, and even they couldn't say for certain what could be expected of the future, indeed whether anything *should* be expected or not. After each treatment Raabi'a's mother-in-law would get her hopes up, and if Raabi'a's period should be a few days late, the old lady would start to suffer from anxiety and palpitations. Immediately she'd vow to perform extra prayers. But when Raabi'a's period finally arrived, her mother-in-law went into swoons of disappointment.

At such times Raabi'a would start to think herself to blame, as though a glass dish had fallen from her hand and shattered. Once she said, 'Amma, my fate is just like that. It's pointless for you to get upset. Why else would my mother have died when I was so young, why would my father just disappear like that, why would I have come to your home with absolutely nothing?'

She wiped away her mother-in-law's tears—who truly loved her, for whom, were it in her power, she would have produced a grandson, *any* grandson—lame, crippled, blind in one eye, stuttering, *any* grandson. But such matters were in the hands of God only, and she had no way of knowing His will, no way of knowing what He might deem to be for the best.

Her father-in-law was a silent observer of these countless trips to hospitals and shrines. He never hesitated giving whatever money was needed, but he also never went on the trips himself with Raabi'a, never put up with the shoving and jostling on the buses. Raabi'a had never seen him argue with her mother-in-law, nor was he one of those fathers who needlessly go on scolding their children. He'd often ask Raabi'a as he was on his way out of the house, 'Bitya, are you sure you don't need anything?' And when she'd respond, 'No, Abba, I'm fine,' he'd follow up: 'Tell me, Bitya. Your mother-in-law is pretty tight-fisted. She squeezes who knows how much money from me in your name; do you ever see any of it? She takes after her father, you know.'

At this the dour veil of sadness would lift from her mother-in-law's face and a soft smile would creep to her lips. Raabi'a enjoyed the teasing that went on between the old couple. Occasionally she would think, 'Did my parents ever treat each other like this? Suppose not ...' The thought made her tired and sad. But then she herself had to smile at these feelings: 'I really am the limit! Look at me, passing judgment on parents I've never even seen.'

She had seen her father-in-law's anger rain down on her mother-in-law only once during her final days. After one last-ditch effort, for which she had taken Raabi'a to a shrine very far away, when she saw that Raabi'a was not praying, her heart began to race. 'What's the matter with you? Aren't you going to pray?!'

At first Raabi'a remained hunched over the rice she was cleaning, but when she did raise her head, tears were shining in her eyes. Out of control, her mother-in-law began beating her. 'Damn you! Everything I've done to help you, and you're going to ruin it all!'

She had no idea where her hands were landing. The tray fell from Raabi'a's hand with a crash, and rice scattered everywhere.

Her father-in-law came out of his room and shrieked, 'Have you gone insane?! Have you lost your senses?!'

The old woman beat her head and cried, '*I've* gone insane? *Me*? Ask this shameful thing over here!'

He grabbed her and pulled her away. 'Are you completely shameless? Raising your hand to the poor girl, no mother, no father....'

Raabi'a had never witnessed him raise his voice like that. Three times she beat her head hard against the wall: she could not comprehend what was going on around her.

She then spoke to her father-in-law. 'Why do you say anything to her? The fault is all mine.'

The old woman put her hand on her heart and began to cry. She moved toward Raabi'a. The old man wanted to stop her, but she eluded his grasp. She took Raabi'a in her arms and hugged her tightly, then began to weep. She was sobbing. 'What...happened...to me?...Why did...you...beat your head?'

Pressed deeply into the old woman's chest, Raabi'a was racked with her own sobs.

Her mother-in-law said, 'I was the one who went crazy. You didn't have to be. May God see fit to break my hands!'

Her father-in-law turned away and headed back to his room.

That evening Raabi'a's mother-in-law said to her, 'Daughter, today I feel like having some *daal-bhari roti*.'

Before entering the delirium that preceded her death, she said to Raabi'a, 'Daughter, I haven't made you miserable, have I?'

She replied, 'No, Amma, you haven't.'

'You've forgiven me in your heart for that day, haven't you?'

She placed her head on her mother-in-law's chest. 'The things you say.'

For a long time she went on talking about her deceased daughter, the mere mention of whom earlier would have brought her to the verge of tears. 'I'm going to meet her very soon. Just look, her husband's family kept her children all to themselves, didn't even let them come visit me. Now she's going to ask me how they are. What will I tell her?'

Raabi'a sat by her and cried. Her husband was standing a short distance away. He was a feeble, emaciated man, and couldn't find it within himself to approach his mother.

He lived on for barely two years after her mother-in-law passed away, leaving Raabi'a and her father-in-law alone in the house.

On both occasions there were multitudes of people who came to express their condolences, including relatives whom Raabi'a had never seen before. The customary rites as well were performed—she not only had to sit with each of them and recite the first chapter of the Koran, but also had to show them great hospitality.

But rather more people came at the death of her husband than at the death of her mother-in-law. Her sister-in-law's children came too, and they all stayed until the end of the forty-day mourning period. Each visitor seemed to be overflowing with sympathy and affection—especially her sister-in-law's husband, who wanted to spend most of his time in the company of his (and Raabi'a's) father-in-law, but from whom the old man always managed to slip away.

Once or twice her father-in-law conferred with her in private. 'Go easy on the expenditure, daughter. All this food, these big vats cooking outside, none of it will reach the dead. I say my prayers only in my heart, you say yours there too. The hearts of those who lift their hands to pray in front of everyone are somewhere else. Their hearts turn to God only when it's time to pray for themselves.'

'I understand, Abba.'

Past the fortieth day of mourning no one had any excuse to stay on any longer. Raabi'a's brother-in-law made known his intention to leave his children there for a while, saying it would do Abba's heart some good. Raabi'a was taken aback when the old man reacted with total indifference: 'No, I cannot let my grandchildren's education suffer just because I'm lonely.'

After all the visitors had left, in truth, utter silence descended on the house—only now, after there had already been a continuous stream of those who came to pay their condolences. Photographs that served only to sadden were taken down and put away; the two spoke to each other from behind artificial smiles. Always small talk, never any mention of the dead.

One morning while sipping his tea Raabi'a's father-in-law asked her, 'What kind of man do you think I am?'

'Excuse me?' she replied in a questioning tone.

'A good man or a bad one?'

'Why? First tell me what you mean, and then I'll answer,' Raabi'a said.

'First nothing, just answer my question!' he said like a child, getting her to smile.

'A very good man,' Raabi'a said.

'But that day I didn't let the grandchildren stay on, you thought to yourself, "What kind of grandfather is this? They want to stay, and he's telling them to go." Isn't that what you thought? Tell the truth, now.'

Thinking it over, Raabi'a said, 'No, I know what was in your heart.'

'Tell me.'

'She died with a broken heart for not seeing her grandchildren. So what good would it do to leave them here now? The more you'd see them, the worse you'd feel.'

'No. Either you really don't get it, or you do know, but just don't want to say.'

'Then what is it?'

'His Highness's eyes are set on this little house. He came here to put his claim on it.'

Raabi'a remained silent.

'And I just can't stand seeing him here. This house is yours alone.' Pausing, then stumbling over his words, he began: 'Did he ever...I mean, did any of the women who came with him...mention...?' But seeing Raabi'a assume her guard, he became too flustered to continue.

From that day forward the workings of the old man's mind began to operate like the parts of a clock. A clock that went on running whether he was asleep or awake. A clock that, every so often, ran even faster, goaded on by a new fear in the old man's mind, a fear born when he tried to put that question to his daughter-in-law, when he couldn't go on.

While his son was alive the old man would go out for walks in the evening. But slowly this too came to an end.

As Raabi'a looked on he began to eat less and less. It was becoming his nature to laugh for no reason—a laugh that masked the absence of any happiness. Many times, after serving him his evening tea, she'd bring his walking stick and say, 'Why don't you go for your walk? I'll serve dinner when you get back.'

But he always said the same thing: 'That's okay. Don't make much food this evening. I'm keeping a fast tonight. The reward is twice as great.'

Silly as it was, she refused to laugh at his excuse. Occasionally she'd mention to one of the neighbourhood women, 'I'm really worried about Abba not eating. If he'd just start eating, everything else would be fine.' She tried not to think beyond this.

As soon as her sister-in-law's husband left he wrote three letters in quick succession, but without even reading them through her father-in-law gave them to her, saying, 'Burn these.'

'You're not going to respond?'

'This is my response. It'll reach him.' As though old age had robbed him of his natural mildness.

He began speaking to Raabi'a in tones that made it seem he was feeling about, trying to discover what lay in her heart.

Inside his head, all day long, the tick-tick-tick-ticking of the clock kept asking him: 'What will become of her? What will become of her?'

One day some women came by whom Raabi'a had never seen before. In order to show them the proper hospitality, her father-in-law had to make an unscheduled trip to the market. The women told her that they had come from the home of one of her father-in-law's friends, and began a barrage of exploratory questions: What was her father's profession? What was his caste? What was her mother's caste?—not her mother-in-law's, but her real mother's?

It was a bizarre situation. She couldn't understand how she was to conduct herself. Should she be courteous to them, as a host? Or should she be shy, like a prospective bride? Her father-in-law was even more bashful—he went to his room, put his walking stick down, and sat on the edge of his bed as though he were just about to get up and go somewhere. Silently she placed the tea and sweets before the visitors, and then came away. Meanwhile, the two neighbourhood women had shown up whom the old man had sent for to sit with the visitors; they were sitting in the courtyard answering all the visitors' silly questions. But Raabi'a looked put out with them too, as though she were displeased with this pointless defence.

That day at lunch nothing was said between the two of them, especially no mention of the visitors—who they were, or why they had come.

The days passed by, and the old man's anxiety only grew worse. From the way he talked one gathered that he felt death approaching, somehow, any day now, but he couldn't bring himself to prepare. Like a man whose train is ready to depart, but who can't get all his luggage in order. Any moment now, in the distance, the train whistle would sound—time for the hands tying his luggage to start shaking. Several times he asked Raabi'a about her relatives: 'Any trustworthy men among them? Anyone I can rely on?'

Playing dumb she said, 'What do you need a trustworthy man for? You have me. Can't you rely on me? I can even do the outside work, if need be.'

'Such as?'

'Fetching your pension. Going to the bank. Shopping.'

Just the answer he didn't want to hear.

Then, one day, he was stricken with fever; shivering, he fell into delirium.

In all these years, Raabi'a was feeling for the first time a fear she had since childhood remained unacquainted with, a sense of her own helplessness, like a plant that had never before sprung from this feeble soil.

After he came back to his senses, he was full of questions: 'All these days I was ill, who stopped by? Who did the shopping? Any letters?'

But then, apprehensively, he said, 'Daughter, I'm a sinful man.'

She remained silent.

'I did you a great cruelty bringing you to this house.'

Choking back her tears Raabi'a said, 'Since when were you the one to bring me to this house? Amma brought me. If I hadn't come, well …'

The old man motioned for her to sit down. She didn't finish.
Presently the old man said, 'Yes, your Amma brought you,
but only with my permission, knowing fully well...You know
Aziz had a wife before, don't you? Yes or no. You know, don't
you?'

'Yes,' she said, barely audible.

'Aziz was a good boy—yes or no.'

'Yes.'

'You know the rest?'

She kept quiet.

'Come on now, tell me. Yes or no?'

'Yes.'

'Yes what?'

'Yes, I know the rest. But you should lie down and get some
rest. Catch your breath.'

'Aziz was a good boy. And I'm not saying so just because he
was my son. He really was very attentive, very obedient. He
never troubled me for money. Nor did I ever hear anybody
complain about him, not in school, not outside the house. When
he was working, he was always so keen to bring me his
paycheck,' he laughed, 'and I said, "If you want to keep it safe,
there are two places: first, the bank; second, the safe your mother
keeps hidden away. I never saw a *paisa* come out of that thing,
only money going in. If you give it to me to keep, it'll just
disappear." He didn't ever trouble you, did he?'

'No...never,' Raabi'a said, sinking into an ocean of sorrow.

'He never gave his first wife any cause for complaint, either.
Still, no two women are ever alike. She couldn't make a go of
it, but you showed how it could be done. Was it because you
liked it here? Or did you feel you had to?'

Though tempered with affection, Raabi'a showed her
annoyance for the first time. 'What are you talking about? Was
I some kind of servant girl, who stayed whether I wanted to or
not?'

'Okay. Now I'm satisfied. But you know, all this time I've
seen myself as someone who's sinned against you. Maybe it
was Aziz's love that did it to me. After all, he was my son.

Even though his heart wasn't really...' The old man checked himself.

'Your mother-in-law started acting crazy after Raashida passed away. Her in-laws wouldn't allow her children to come here. Poor woman, she wanted nothing more than to raise her grandchildren herself.

'The old lady knew her mind,' he said with a smiling voice, as though his dead wife were standing before him. 'She had Aziz remarried ahead of time, hoping that if a daughter-in-law were around, she'd look after the children. I just couldn't argue with her. It was pointless to expect rational behaviour, especially from a woman whose daughter had just died in the prime of her youth.

'So her second big blow was not seeing her grandchildren, and her third was watching Aziz's wife go back to her parents' house.

'But then she found you.' He laughed for a bit. 'Maybe she found you in order to take care of me in my old age. I can just hear her: "After I'm gone, *somebody* should be around to drip a few drops of water down the old man's throat." Oh, the games we play in life. So I might not be the sole guilty party; still, I certainly had a hand in it.

'A time comes in a man's life that tests him. Now I think that I should have held up then. Should have told myself, "Look, such was our fate, period."'

He thought silently for a while, then suddenly spoke up: 'But now I'm about to ruin all her plans for this house.'

It seemed to Raabi'a that she and her father-in-law had already climbed aboard two separate boats. One was pointed toward the far shore, behind which lay only jungle. Someone had come and untied the other from its moorings, but had not climbed aboard with its traveller.... That unsteered boat was now flowing along with the waves, gradually fading from view, into obscurity, out toward the sea. She wanted to cry out to those on shore, 'Somebody jump in! Somebody stop this boat!' But she was unable to utter a sound.

When she awoke her father-in-law was in a chair trying to adjust the small cushion behind his head. She was drenched in perspiration.

'Go back to sleep,' he said.

It was as though Raabi'a had found some peace. Her eyes, awake the whole night, closed again.

A few days later, when Raabi'a brought her father-in-law a bowl of broth, she found him lying on his bed cleaning his glasses with the edge of his *kurta*. He said to her, 'Have a seat.'

Raabi'a sat down.

'You know what your mother-in-law had in mind when she chose you, don't you?'

'Yes—um, no,' she said without much thought, soaking up the breeze from the fan.

The old gentleman chuckled and said, 'You're becoming quite the politician—give two responses, and one is bound to be correct!

'She didn't look just at your appearance, nor was she simply satisfied with your household skills. She was completely taken with your name. She told me, "*Raabi'a* and *Raashida* sound so much alike," and then she broke down crying.'

Silently then, he began sipping spoonfuls of the broth.

When Raabi'a came to retrieve the bowl, he again said to her, 'Have a seat.'

For some time she sat quietly, waiting apprehensively to hear what the old man had to say.

He spoke up. 'When I was sick, I had the strangest dream. I forgot to mention it then. It came to mind today, so I thought I'd tell you about it.

'I dreamed it was evening in this house. As though someone had passed away here, and one by one all the mourners had already left.

'And after the last person had gone you locked the door from inside and were completely alone in the house ...' His voice was ragged.

Raabi'a wanted to ask him, 'Are you telling me of your dream, or of your fears?' But she was swept away by the deluge of her own thoughts.

As though he had finally emerged victorious from all this, as though the moment had arrived at last for which he had been preparing her all this time: after the muted ceremonies one sees only when a widow remarries, somber-looking members of the groom's family silently escort her to her new home after the wedding...Then, after the last of the guests leave—just a few people from the neighbourhood, really—the old man bolts the courtyard door from the inside, returning with exhausted steps to his room in this frightful, lonely house....

The old man asked, 'What is it?'

'Nothing,' she said, as if awaking from sleep. She thought she should perhaps tell him her dream, too, but she felt a wave of compassion for the old man wash over her, so she didn't.

The thought occurred to him: 'What good is it even to think about these things? It's hardly likely that she'd get married just because I said so. I've been frightening the poor thing for no reason.'

A breeze of contentment passed through the room, gently touching them both.

—Translated by G.A. Chaussée

The Poor Dears

A cold, wet wind was blowing outside as my plane landed at Heathrow Airport. In the terminal building I spotted Cathy among the crowd coming to receive their relatives and felt reassured. I was no longer worried about how I would get to my flat.

I dumped my baggage in the boot of Cathy's white Ford and flopped down beside her on the seat. After she pulled the car out of the airport traffic into a calmer street, I lit a cigar and said, 'Open the window a bit, or you'll wind up dead behind the wheel in your quiet, cozy world.'

'You may keep smoking,' she said, without taking her eyes off the road.

I closed my eyes and dozed off. I awakened only when I felt Cathy's hand trying to remove the cigar butt from between my lips.

'How did you know I was asleep?' I asked.

'How? You snore.'

We had reached my flat. Standing on the dark, cobbled street, I thanked Cathy, promising to tell her about my travels in the morning, and said good-bye to her. Then I looked around impassively. Empty milk bottles stood outside doors in the dim, grey light. Except for one, all the other flats were dark behind curtains that had been drawn shut.

The same old place. I was home for certain! It is a rare virtue in Cathy that she never asks questions unless I am in a mood to talk.

In the morning I went through my mail. There were bills, bank statements—the usual stuff.

The cleaning lady came and started work. At one point she asked, 'How did your trip go, Mr Hasan?'

'It went quite well, thank you,' I said, offering her the red box of Benson & Hedges.

She thanked me and took out a cigarette. As she put it carefully away in her apron pocket, she said, 'I mustn't waste such an expensive cigarette. I'll smoke it after I'm finished with the cleaning.'

I gave her the whole box and said, 'Come on, old girl, have a smoke with me first. You can always work later.'

'So you felt quite at home there, Mr Hasan?' she asked producing a cloud of smoke.

I told her that if by 'there' she meant India and Pakistan, then she was mistaken. I hardly even knew the names of my relatives in those places. Or if she meant some other countries, then, until a few months ago, I knew no more about them than did any ordinary Londoner.

I scarcely remember when the cleaning lady left and when Cathy walked in. She had come to take down my travel notes, which she would then type and return to me properly arranged, so that I could start work on my new book. A number of trifling chores which, like any ordinary traveller, I had foisted upon myself, kept getting in the way, however. I wanted to get them out of the way. For instance, the decrepit old man I had run into at the Angkor Wat ruins—the grand temple dedicated to Lord Vishnu—had asked me through a bout of nasty coughing, 'I hear that in England they have come up with a brand new drug for curing bronchial asthma.'

Nearly exhausted from my strolls through the myriad pathways and balconies of the temple that sprawled over some five hundred acres, I had just sat down, removed my burning feet from my shoes, and plunged them into the cool, refreshing dirt. The old man sat close by, carefully holding my movie and still cameras in his lap to protect them from the dust. In my tranquil surroundings, I began to make notes.

The sky was crowded with dark, low-hanging rain clouds. The old man panted for breath. 'I hear,' he said, as if to himself, '*that* drug roots out the disease.'

I jotted down the old man's name and address in my notes on the Angkor Wat ruins, setting them apart carefully within parentheses. Even as I did that, I couldn't suppress a smile thinking how Cathy would type out the lines on a separate sheet of paper which she would then hand over to me saying, 'Perhaps this gentleman belongs to the present century; we can't possibly include his name and address in the book.'

I promised the old man to ask my doctor friends back in England about the drug. If such a drug really existed, I would be sure to send it to him. One of my journalist friends could bring enough of it to last him six to twelve months.

I had been expecting my offer to brighten up the old man— amazing, isn't it, how the East eagerly awaits every new discovery or invention to come from the West!—but his face remained entirely expressionless. With measured politeness, he said, 'All right, sir, if you say so.'

The old man handed me back my cameras and got busy in his work. He had as little hope of hearing from me again as he had of being able to fall asleep that night; and failure to get the drug would have hurt him as little, I thought, as the realization that his life was coming to an end. He would have accepted them both with the ascetic detachment befitting a Buddhist monk. 'Time' and 'human dependence' had become meaningless words for him in the ruins of that centuries-old temple, which resembled a veritable town in its vast sweep.

Then there was this other *bhikshu* I had met in a Buddhist monastery in Sri Lanka. I had mentioned to him a recent publication from New York and London which contained colour pictures of all the important Buddhist monuments the world over.

This monastery, flanked on all sides by king coconut palms, was a short distance from Colombo. I was visiting it for the second time. I had first come here when I was on my way to Kandy and Anuradhapura, and now again as I was leaving Sri Lanka for India. In between, I had managed a trip to the ruins of Anuradhapura and taken pictures of the brick *maths* erected during the Sinhalese period in honour of Buddha.

The Temple of Buddha's Tooth at Kandy, with the delicate art work on every stone and beam, was also fresh in my memory. I was terribly disappointed that they would not open for me the room which housed the celebrated relic, so that I could verify for myself whether there really was a tooth or whether the gullible people had just placed their faith in a velvet box.

The ceiling of the balcony outside this room held a painting of an elephant. I would have passed through the balcony almost without noticing it, or would have at most given it a cursory glance, had not one of the temple devotees invited me to look at it carefully. I looked up without enthusiasm, hoping to find a profile of Ganesh, the elephant god, but stopped short, and the devotee let out a short, gentle laugh, which I am sure he must have laughed many times before. The accomplished painter had so skillfully depicted a bevy of young female shrine-devotees that they merged into the figure of a colossal elephant. I asked him about the painter but he chose to ignore my question, thinking, perhaps, that it was insignificant. Instead, he started telling me about some white flowers that lay in front of the statue of the seated Buddha.

I was to ask the same question of my *bhikshu* friend in Colombo during my second meeting with him. The intervening period, which I had spent in different cities of Sri Lanka, had created a sort of closeness between us—a closeness which was obvious from his expression but which he was loath to admit. I guess this was because intimacy bred the very attachment which he had wanted to curb by renouncing the world and adopting the austere, saffron-dyed garb of a *bhikshu*. My single question prompted him to ask a few of his own about India, about Burma. Finally, I gulped down the coconut milk he had offered and got up.

It was to him that I had promised to send the book. I telephoned Foyles. They did have it in stock. Price: two pounds. I placed the order and instructed them to send it directly to the monk in Sri Lanka. This done, I crossed his name off my check list. I felt a sense of relief slowly coming over me—I couldn't have wanted it more.

A little while later Faiq Ali telephoned from Manchester. He wanted to know about his relatives whom I had met in Karachi. It was strange, wasn't it, that before embarking on my voyage to the East it was I who had asked him, 'Well, aren't you going to give me the addresses of all those first and second cousins you keep telling me about all the time?' and now it was he who was so impatient to find out from me about those same 'first' and 'second' cousins: 'What kind of people are they? Did they treat you well?' So on and so forth, as if Naima and her relatives were in fact mine, not his.

The only reason I had asked Faiq for addresses was so I could see first-hand how people lived in Pakistan, what sort of problems they had, what kind of hopes and dreams they cherished. One rarely gets an accurate idea of a country and its people by putting up in hotels, or reaching out to them through tourist guides and travel books.

The morning after I arrived in Karachi from Colombo I first confirmed my reservations in the hotels where I was to stay during my visit to different cities in West Pakistan, then made a few phone calls regarding my schedule. Finally, I called the European drug manufacturing company where Naima worked. I had thought everybody would know her there. This was not the case. I was told that she was just a packing girl, free to talk on the phone only during the lunch hour. I had not finished talking when the receptionist rudely hung up. I dialed again, this time asking to speak with the General Manager of the company. The man at the other end sounded irritated. Why didn't I go to Naima's house and ask her whatever important thing it was that I wanted to ask her? Why was I wasting his time? But when I told him that I was a writer from England on a trip to Pakistan and knew next to nothing about this country, his voice changed noticeably. He asked me courteously for my phone number and instructed someone on the intercom in Urdu, 'Look, there is some girl called Naima who works in the packing department. Ask her to come right away to my office and take the phone.' He then politely asked me to wait awhile.

As I waited I could hear faint snatches of some Pakistani music playing in his office and two men talking in one of the regional languages.

When I picked up the phone again I heard the voice of a frightened female at the other end talking in barely audible tones. This was Naima who, I sensed, was quite embarrassed talking to a perfect stranger like myself in the inhibiting presence of her boss, scared that her voice might ruin the decor of his room. I guessed from her voice that she must have been around twenty years old.

'What do you want?' she asked in a whisper.

The rest was more or less a monologue. Had it not been for her faint 'yes'es that echoed dimly through the receiver from time to time, I would have thought the line had been disconnected, all the more so as the sound of Pakistani music had meanwhile died out and the other human voices, too, had stopped.

As I talked to her I couldn't resist imagining a frightfully pretty girl at the other end—all alone, skewered by the lustful stares of the men in the room, trying her best to crawl into the receiver to avoid those piercing glances, but who did not know how. Even her 'yes'es were no longer audible to me.

I told her that I was a friend of her cousin Faiq and had come to spend a few days in Karachi. Then I asked her if I could, perhaps, come and visit her at her house. I asked her for her address and mentioned that I would drop by some evening after my return from Taxila.

'Tell me the day you want to come and give me your address,' she whispered. 'I will have my brother come and pick you up.'

This was the second and the last full sentence the girl uttered. I fished through my date book and gave her my address and the date on which I was scheduled to leave for Beirut.

I related the details of my meeting with Naima and her folks to Faiq. All he seemed to be interested in, though, was my comment that if he was looking for an Eastern wife he would not find a better girl than Naima, from east of Suez to Cambodia.

To return to my meeting with Naima and her family: On the appointed day, the hotel receptionist sent to my room a youth whose face was all but covered with pock marks. He looked just as frightened as Naima had sounded over the phone. He had come to escort me to their house. I offered him some refreshments but he declined, adding courteously, 'Back home, we are all waiting for you to join us for tea.'

The poor man seemed even more over-awed by the hotel than by me. It was probably the first time that he had set foot in it, and was feeling quite out of place in its plush, swank environment. I asked the waiter for a taxi and came out of the hotel with the youth in tow.

Throughout the ride I remained silent. I didn't wish to embarrass my young companion further. He, on his part, preoccupied himself with giving directions to the cab driver. We passed through different parts of the city, each with its peculiar lifestyle. The faces, bodies, and garb of most of the pedestrians suggested that we were proceeding from affluence to poverty, from a world of plenty to a world of dire need. The women in Naima's neighbourhood flitted about in veils. Children, some barefoot, some with runny noses, romped around. Here and there along the street some people had set up cots on which they sat or lay. There were no foreigners.

Naima's brother led me into a dull, pale building. We climbed several flights of dark, dank stairs and entered a flat on the third floor. I had to spend some time alone in the living room. In fact I had expected that and was mentally prepared for it. As I sat there waiting for my hosts to appear, I realized my mistake. I should have met these people immediately after my arrival in Pakistan, so that on a second visit around the time of my departure we would have become informal enough for me to gauge their true feelings, and to have some idea of their hopes and aspirations. It is amazing how a first meeting, no matter how protracted, almost never creates the same degree of informality as that generated by the interval between two short meetings.

The first to enter the room was Naima's mother: middle-aged, sallow-complexioned, tolerably good-looking—I thought. Next came Naima's sister. She looked more like a younger sister of the middle-aged lady, with nothing striking about her. The last to enter was Naima herself. She was truly stunning.

I had thought I would spend at most an hour with them. But I ended up spending the whole evening. By the time I got up to leave, I had become thoroughly acquainted with the entire family and their past life.

The hospitality started with fried snacks. Later, the older daughter, succumbing to the old lady's persistent requests, sang a Mira *bhajan* for me, and Naima, again at her mother's behest, played a cracked disc on the gramophone, to which I listened with feigned interest. I was also formally introduced to the photographs which hung from the wall. One of these, a picture in copper tones, shot most probably some time between 1930 and 1940 and printed on orthochromatic paper, was of the girls' father. Like the occupants of the house, I, too, had to pick up the picture and look at it in reverent silence for a while before replacing it on the wall.

I promised to write to them and apologized that I must do so in English as I wasn't fully conversant with the Urdu script. Throughout the evening I had noticed how my halting Urdu had amused them. I also promised to have Faiq's mother write to them as well. However, when I got up to leave, I knew deep in my heart that I was neither happy nor satisfied with this meeting. I noticed that everything which might be even the least bit offensive to look at had been deftly removed from the scene. Soon upon entering the room I had spotted items of laundry left to dry on the clothesline on the balcony. But when I got up to greet the lady, my eyes fell accidentally on the clothesline and I was mildly surprised to find it bare. It was as though somebody had in the meantime crawled to the balcony unnoticed and pulled the laundry off the line without attracting attention.

After talking with Faiq I crossed another item off the check list.

The pictures had been sent out to be developed. I expected them back within a few days. I would then send Amand the pictures of his family as well as the baby overall with the zipper which I had promised him.

I had met Amand on a lake in West Pakistan where he worked as an oarsman. He had given me the most detailed information about this region. He had told me how the lake was once the land between two hills and how the waters of the neighbouring river had been diverted to fill it. And, there, on the island that I saw, was the shrine of some venerable woman saint. Formerly people walked all the way to it for pious visitation but now, since only a few could afford to pay for the boat ride, most returned from the waterfront after making their votive offerings.

Amand's family had given me coarse reddish bread of rice flour to eat and a single fish to go with it, which he had borrowed from a fellow oarsman and fried for me. Colour photos were not the only things I had promised Amand. I was going to send him an overall, too, for his baby who was spending the last trimester in its mother's womb. Bundled up in the overall, the baby would be freed from the danger of catching pneumonia from the lake's cold winds.

Most of Amand's children had suffered from acute bronchial pneumonia—I had guessed as much from the description he gave of their illnesses—but he and his family firmly believed their ailment to be the work of some evil spirit, which, in fact, as they thought, had even claimed a couple of Amand's children's lives.

Even in their wildest dreams, Amand and his family couldn't have imagined such an overall, let alone owning a brand new one. This overall was going to be the expression of my gratitude to them for their hospitality and service.

The list began to shrink—slowly, gradually.

In time I crossed off Amand's name, too; as well as that of the old Catholic lady who taught school in India to whom I had mentioned having seen the first resting place of Saint Francis Xavier at Malacca—that rectangular pit from which his body was later exhumed and carried to Goa and reinterred there. 'I'll do anything you want,' the old woman had entreated me most respectfully, 'if you could, perhaps, send me a picture of that pit.'

And I had promised that indeed I most certainly would.

I did some stock-taking of myself after crossing the old lady's name off the list. I had been back in London for a good fortnight now and had started work on my new book. My life had swung back to its normal rhythm, the one it had before I began my travels to the East: reading newspapers, writing, other chores, study, visits to the library, afternoon strolls, then TV and sleep.

One day Cathy picked up the packet which had been lying on my table for the whole week and asked, 'Aren't you going to mail it?'

I was in the other room, so I asked, 'Mail what?'

'This packet, addressed to Miss Naima so-and-so, Karachi?'

I returned to my study and asked her, 'Do you know what's in it?'

'Two hundred feet of Scotch magnetic tape ... made of polyester, right?'

'Is that all?'

'Well, the tape is enclosed in a cardboard box which is wrapped in soft padding. You have had the housemaid wrap the whole thing in cloth and sew it up and then enclose it in this manila envelope.'

'But the tape—what's on the tape?'

'Your message for the girl—isn't that rather obvious?' Cathy said, returning the packet on the table.

'No, I wish it were that simple. For then, I would have either mailed it myself or asked you to mail it for me.'

'Your friend Faik's message, then,' Cathy said, thinking hard, 'or maybe Faik's mother's voice ... for that *sallow-complexioned, good-looking, middle-aged woman*?'

After a while she asked jokingly, 'Which of these two, the girl or her mother, is likely to be the central character in one of your future novels?'

I remained silent.

'Still hung up on mothers, eh?' Cathy continued. 'When are you going to outgrow this fixation? Why not the young lady ...'

But I remained impassive. My expressionless face prompted her to probe, 'Well, aren't you going to answer?'

'Cathy,' I began, 'every single day for the past week I have thought of mailing this packet but have put off doing so for one reason or another. You have no idea whose voice I have taped on it.'

'Well, whose voice?'

I ignored her query and continued, 'I have been thinking all week long whether I should send the packet to Naima. I ask myself, now that I have got the tape and have gone through the trouble of having it neatly packed, why not take it to the post office, have it weighed, put the stamps on it, and mail it? But then I think, suppose I later regretted it, nothing would stop the packet from reaching her. I am finding out, for the first time in my life it seems, that whatever you have committed to another ceases to be yours.'

'What, for instance?'

'For instance the arrow committed to the wind, the dead body to the earth, and ...'

'There you go again,' Cathy interrupted. 'It is the Eastern man inside you that makes you say all this.'

I continued. 'Cathy, I am unable to decide what to do with it. Once or twice I have even run my fingers over the wrapping to see if it has gathered dust and then laughed at the foolishness of my act. There is no dust here. How can there be any in cold countries? Perhaps I am driven to do so by my desire to find out how long it has been lying on my desk. You see, in the East, they judge the length of time passed over a thing by the amount of dust it has collected.'

'Hold on, let me grab a pen and notebook,' Cathy said in dead earnest. 'I guess this must be part of the book you are writing now.'

But I went on. 'Then again, it is entirely possible that I am no longer quite so anxious to send it off to Naima. The book eats up most of my time. As I work on it, I become completely oblivious of Naima, her dead or living family members. Moreover, a busy writer, in search of new materials for his book, soon forgets the people he meets and photographs he takes of them during the course of his travels, and the promises he makes to them. There comes a time when these people, uniquely individual and vibrant with life, are transformed into mere characters, and all the places he has visited become the stage on which all the different acts of the cosmic drama of life are enacted all at once.'

'I can easily use this material for the Foreword of the book, you know,' Cathy said. But I continued in a slow, halting voice: 'It is quite possible that Naima and her family have by now become mere characters to me, and that this packet no more than a mere reminder of the time when I had just returned from my travels in the Pacific and the Indian Ocean, of a time when I had noted in my diary what I had to do or send to whom.

'You remember I had told you how at my request Naima's brother had come to take me to their house, don't you?'

'Yes, I do,' Cathy said, putting aside the pen and the notebook on the table.

'Let me go over that scene once again. Then when you have heard the whole story, tell me whether or not I should send the packet on to her.

'Well, every single object in that living room disguised an overwhelming desire to be recognized, to be esteemed. That is why everything that failed to measure up to their standards, that seemed mean or odd or otherwise betrayed poverty, had been spirited away from the scene. Naima's brother, who suffered from some chest ailment, told me that he worked for the railway. But he never did tell me what exactly his job was. Naima, too, was more than a little diffident about the nature of her own work at the drug company. And time and again the mother kept saying,"You cannot even imagine what their father was and all the things he wanted to do for his children." But when I asked, "Well, what was he?" she answered, "An *artist*!" She also told me that Naima was born after her husband's death. It was at this point that they removed that picture from the wall and showed it to me—the picture of a young man shot on an orthochromatic plate, who couldn't have been more than thirty years old, I thought, when he left her widowed.

'My earlier excitement at meeting the attractive widow had begun to wane in that stuffy, lackluster atmosphere. I was looking at everything without enthusiasm or interest. I was told that it was Naima's elder sister who had inherited all the artistic talent of the girls' father—perhaps because she was fortunate enough to have been raised by him—for she sang very well and was an accomplished vocalist, while Naima, well, let's just say she had been trained for a career job right from the start.

'I was at a loss. I had no idea what they took me for. Had my movie camera led them to believe that I could, perhaps, put the girls in films? The accolades they liberally showered upon the older girl forced me to ask her for a song. Not unexpectedly, she declined. Ultimately, giving into the persistent, urgent pleas of her mother and younger sister, she did however sing a *bhajan*, by Mira Bai, I was later told. I was now beginning to pity them. The girl, or woman if you will, simply couldn't carry the higher notes of the song.

'Next they bragged about their record collection. Some of the discs had been collected by the girls' father and some, after his premature death, by their mother. One by one they showed off every single disc. As I looked at them I couldn't help feeling they were light years away from the age of 33.3 and 45 rpm records. They were all old 78 rpm discs—bulky and awkward, which you played by changing the gramophone needles every so often. Their center labels—depicting the yellow Gemini Twins, an elephant trunk, a lion, a horse—were a novelty to me and I wanted to buy a few of these relics and bring them along. I had seen them being sold, along with used books, in Sadar, the city's biggest shopping center. Their grooves were all but gone, some didn't even have any grooves left.

'Naima's brother sat silently on one side. Outside the window daylight had waned. The middle-aged woman handed me a record and said, "Here, this is their father." I looked at the record. It had a light blue label with the picture of a pair of spotted deer and the legend: Calcutta Recording Company; Music by: R.C.B.; Orchestra conducted by: P.K. The remainder was in Hindi. The record had a hairline crack running all the way from the center to the outer rim where it had been deftly mended by a piece of copper sheet and minuscule nails.

'I asked them to play the record for me and they were quick to oblige. The voice sounded vaguely familiar. I knew I had heard that song in England at the house of one buff or another of old Indian music, but try hard as I might I failed to recollect exactly where. Just then I heard the lady say, "Watch, here he comes!" What came was a piece of flute music. Naima and her sister pointed simultaneously at the rotating disc and screamed, "There he is!"

'The record kept playing, making a click each time the needle hit the crack. Once it even got stuck in a groove so that Naima had to quickly lift the heavy playing head and advance it a couple of grooves. I was trying hard to avoid looking directly into their faces. The flute intermezzo was incorporated several times in the composition and each time it was played they listened to it in hushed reverence, while I wondered, what if the record broke one day ... what would they hang on to, just what?

'On my way back to the hotel I realized with frightening clarity how terribly incomplete all my notes were. The most they could do was tell me the names of the kings who had built those temples at Anuradhapura and Kandy. Granted, the accounts of the monuments at Delhi and Agra were somewhat more detailed; for instance, it was possible to find out who had designed a particular building, who the architect was; but could they tell me, would I ever know anything about the man who had actually picked a particular stone or slab and carried it there, or about him whose dexterous hands had wrought such marvel on that stone?'

'Well, let me tell you what to do,' Cathy said. 'Just let this two hundred feet of magnetic tape sit right where it is. I think I know what it is that you want to send to that family. But are you sure they can afford a tape player to listen to it?'

'I can easily send them one,' I said in a choking voice.

'And make them realize that all you noticed about them was their poverty? That besides that snatch of flute music, to which they clung so miserably, they owned absolutely nothing? Surely you don't want to insult them, do you?'

Cathy picked up the packet, played with it for a few seconds and put it back down. Then she said, 'Just let it lie here. With it before your eyes as you write, the individuals you wish to talk about will not turn into mere characters.'

After a brief silence I asked, 'Care for a walk?'

'Sure, why not. To my place?'

—*Translated by Muhammad Umar Memon*

White Man's World

In those days many things had gathered in my mind. They are there even now. I think the minds of all of us children were full of inquiries. We would question each other about some of the things, or would approach the grown-ups. Mostly the grown-ups didn't have time to listen to us. Even when they did answer our questions, they alone seemed satisfied with the explanations they gave us; we knew that they didn't know the real answers.

Take, for example, this question: When a person dies and we dig a deep hole in the ground and leave him there under a pile of earth, who comes to him?

Angels?

But I knew that neither I nor any of the grown-ups had ever seen any angels. And, then, when someone was buried under a pile of dirt and shut tightly inside a wooden box, how could any angel get inside to talk to him and carry him to the Heavenly Father? These were all lies.

My older sister is very fond of reading. She lies on her stomach all the time to read her story-books. When I am on speaking terms with her, she sometimes reads a story to me and to my younger sister. She has a whole library of books. Now I, too, have a few books. When we have a house of our own—Mommy and Papa say we're going to have one, one day—then I, too, shall have my own library.

Most of the books my sister has are children's detective novels. Or they are stories of four people, one of whom is always a woman, who go on a mission, assigned to them by a person or a government, and, in the end, come back home after blowing up a big building or a bridge.

What the word 'government' meant I did not know then. I'm still not quite sure. It was something that was for the grown-ups, not for the children. It was the thing that transferred Papa

from one place to another and did many other things, which I
shall talk about later.

So, as I was saying, my sister has quite a few books; of these
many are by Enid Blyton. Sometimes my sister and Mommy
have an argument. My sister says that no one can write a better
story than Enid Blyton, and that the greatest book in the world
is *Heidi*, but Mommy says that there are many great writers in
the world—Tolstoy, Thomas Mann, and Dostoviskido ... I could
never get that name right. Mommy once told me a story by that
man, and I thought it was a good story. There was an old woman
in that story who was very miserly, and a student who never
had enough money to eat and who used to sell some things to
that old woman. One day the student made a plan and murdered
her, and nobody could find out who had killed her. In the end
the boy himself went and told everybody that he was the
murderer of the old woman. The story was called *Crime and
Punishment*. The old woman in the story reminds me of my
grandmother who lives in Pretoria in South Africa.

I hate my grandmother. She lives all alone in a house which
is in an area meant only for the houses of the rich. She proudly
says that only the whites live in her area, no Asians or Africans.

That is why I cannot stand her. Besides, she has hardly ever
given me anything. Once, long ago, when we visited Pretoria,
she gave me a battery-operated jet plane. The jet used to make
the kind of noise planes make when they take off. On its
starboard side it had a green light and on the port side a red one,
perhaps. It would move on the ground and, after a while, make
a breaking sound and then stop.

This goes far back in the past when I was really small. But
even now, when we go to see her, she reminds me of that gift,
and says, 'Remember, young man, the beautiful jet plane I gave
you? Where is it now? Does it still work?'

On a wink from Papa I have to say, 'Yes, Grandma, I still
have that plane and it still works beautifully.' Actually, I've
even forgotten what it looked like and what side of it had which
light.

The old lady always asks my mother, 'Where are you people finally going to settle?' I know the meaning of the word 'settle.' It means to build a house somewhere and to stay there forever, as the old lady has done herself. Then she asks, 'Are you going to Australia? Why not Canada? Hank could even find a place for himself in the States.'

Hank is my father's name.

My grandfather was Dutch and grandmother Afrikaner. Perhaps they hated each other all their life. My grandfather spent the greater part of his life in Indonesia, just so he would be away from my grandmother. I've heard Mommy and Papa whisper that my grandfather, later in his life, had started living with a woman who was not white. This was one of those things which stayed in my mind in those days and which raised a number of questions. Why did my grandfather do that and, if he did, what was wrong with it? I had also heard that the government did not allow him and that woman to live in South Africa, and that once, in Durban, when they were travelling together in a taxi, they were taken to a police station because the two of them had different skin colours, and the government didn't like that. Things are not that way in this country where we live. Here, a Negro woman and an Indian man, or a Negro man and a British woman can go together wherever they want to—to the movies, to the circus, or just for a stroll, holding each other's hands, as Mommy and Papa often do before sunset. I like to watch them when they do that. They seem to be in love with each other. When I grow up and get married, I too would sometimes like to go out with my wife in the same way. My wife's hand would be around my back, as Mommy's is around Papa's, and my hand will go behind her neck and rest on her far shoulder. But what worries me is that if she is taller than me, then walking like that may not be possible.

Anyway, there is a government in the South—Mommy and Papa call South Africa 'South'—as well as here in this country where we live.

Once I remember seeing my grandmother panting and breathing hard, as if she were trying to hold back her anger, and saying to Papa, 'I don't care a hoot where your father is, or if he is dead or alive. He was low and that is what he proved by living with a dark woman. Sin or no sin, the important thing is that he lived with a dark woman.'

I knew what the word 'dark' meant: people are afraid of the dark and of things that are dark like the night. In the school plays I've seen the devil wearing black clothes and black make-up. This continent, where we live, used to be called the Dark Continent. I imagine that long ago people used to live in big caves, and since there was no electricity, they were afraid of coming out at night, as people still are in this small town where we live and where there is no electricity. People don't walk around at night because they are afraid of snakes and big scorpions, some as big as my two hands put together, and of lions and hyenas. But me, I'm not afraid of the dark, nor do I fear black people.

One day when my father and I were going for a morning stroll, with our dog Mishka in tow, I asked him, 'Papa, why don't we live in the South?' Papa knocked down some *neem* fruit for me from a tree with his walking stick and said, 'Do you want to live there?' 'No,' I replied, 'but even then, ...' and started nibbling at my *neem* fruit quietly. Mishka sniffed the bitter *neem* berries and then went into the bush after some animal, perhaps a jungle rat.

Once again I asked, 'Papa, is Grandpa really a low man?'

'No,' he answered, so firmly that I was convinced that he really meant what he said. 'No, your grandpa was not low; in fact, he was a great man.' Then, after a pause, he added, 'I say "was" because he has now passed away.'

I couldn't quite figure out what to say to a grown-up at such a time. Obviously, I couldn't pat my father on the head, as he does when I'm sad or when Tina or Fiona is; nor could I ask him to bend down so that I might pat him. I couldn't even say to him, 'Please don't feel sad.'

It was a beautiful morning. Mishka had once again joined me, and the sun had just come up on our right from behind the sheet of mist.

Suddenly I asked, 'Papa, why did the government not allow Grandpa to live in the South with that other woman?'

In embarrassment he answered, 'Perhaps there wasn't enough room for more people there.'

I knew he was lying because at night Mommy would often read an ad from her medical journal and tell Papa: 'Here's another one of those ads: emigrating to South Africa?' This journal was published in England, and I had heard that ad many times: IF YOU ARE LEAVING FOR SOUTH AFRICA, CANADA, AUSTRALIA, OR THE UNITED STATES, WE CAN LOOK AFTER THE PACKING AND SHIPPING OF YOUR HOUSEHOLD EFFECTS.

It seems as if the whole world is being invited to settle down in South Africa, but there is no room there for my grandpa and a dark woman.

This situation is like a private joke with our family. Every house has some bits of stories whose mention amuses every member of the household. In our house this is what makes both Mommy and Papa laugh. Little Fiona laughs, too, although she is so stupid and doesn't even remember meeting our grandmother. On such occasions Papa utters a word: 'hypocrites.'

When I asked Mommy about that word, she said it is used for a person whose beliefs and acts do not match.

But Fiona still regards hypocrite to be a big animal like a hippo, or even bigger than a hippo, like the animals we often see in the rivers here, who sometimes overturn boats.

So, that morning, when I, Papa, and Mishka were returning from our walk, I asked him, 'Papa, if I grow up and marry a Negro or a dark woman, would it be something bad?'

Papa laughed and said, 'No, it won't be bad. Do you like dark girls?'

I nodded.

Then teasingly he asked, 'So when are you getting married?'

'Not right now,' I answered, and both of us laughed.

How we came to this small town is yet another interesting story. Papa had been transferred here from Sierra Leone. Upon coming to this country we stayed for a few days in a hotel in the

Capital. As South Africa is simply known as the South, the main city here is called the Capital. When this word is used, everyone knows what city is being talked about.

Then when the government ordered Papa to start working in this small town, we went to the Capital's railway station. The place looked deserted, and a big carriage which had once been used by some members of the British royalty stood in the middle of the waiting room. A small metal plaque with some inscription on it had been fixed in a cement block nearby.

Tina read the inscription and said that during some year in the 1800s an English king and a queen had visited the city in this carriage.

I heard Mommy say to Papa, 'They should do away with these memorabilia.'

I didn't quite understand her last word, but I didn't want to ask Tina for fear she would begin putting on airs.

Later, when we boarded the train, we met a white woman who was already seated in the compartment. She looked at Mommy, heaved a sigh of relief, and said to the two men who had come to see her off, 'Thank God. I think I will be able to sleep in peace.'

Tina picked up her book of stories and went up to the upper berth. Fiona started rubbing her nose against the window pane and watching outside, and I waited for the train to move so that I might start talking to Mommy and Papa. In the noise of the train I would be able to talk freely, and that woman wouldn't hear our conversation.

I knew well what Tina was reading. Most of the stories she used to read those days I've read by now; the rest she has told me about. That day she was reading stories by Tolstoy. One of these stories was about a farmer named Pakhom who had a little land and was happy. But then he became greedy and started buying more and more land in order to become rich. At last he decided to go to the country of the Bashkirs where land could be had for free. There was only one condition: after leaving in the morning he would have to return before sunset to the place from which he started; all the land that he walked around would become his. But the devil put greed in Pakhom's heart, and he decided to circle a huge piece of land. He walked so much during the day that when he returned to the Chieftain of the Bashkirs, the sun was already setting. He fell on the ground as he came near the Chief. The Chief said, 'Now here is a real man. See how much land he has acquired!' But when Pakhom's servant tried to lift him up Pakhom spat blood and died. So the servant had to dig a grave and bury him. At this point in the story, Tina would declaim, 'From the top of his head to the soles of his feet, all the land he really needed was six feet.'

Tina would say this as if she were delivering the Sunday sermon from the pulpit, and as if I were someone greedy like Pakhom who had to bow his head and listen humbly. The story was called 'How Much Land Does a Man Require?'

I had made up my mind long ago to read the story myself, for I felt that Tina twisted the facts, or that, perhaps, she had in her the makings of a church-going, religious woman. I would have to find out for myself how much land a man really needs.

The train was about to leave when an Asian woman, almost at a run, holding a baby in her arms, approached our compartment. Behind her was her Negro servant who was carrying an older child in his arms. A girl almost my age walked into the compartment, all the while talking to the coolie.

On seeing this crowd, the white woman at first panicked and said 'Oh, no!' But when she heard the Asian woman speak to the coolie in English, she calmed down a bit.

The coolie placed the luggage in the compartment and left. The Asian woman and her family occupied one full berth. The Negro servant also sat on one edge of the berth.

The train whistled.

The white woman, confused and angry, pointed towards the Negro servant and asked the Asian lady, 'He is not going to stay in this compartment, is he?'

The Asian woman answered her in English, 'His seat is also reserved in this compartment.'

'No,' she said sternly. 'He is not to travel in this compartment.'

She called the Negro guard who happened to be standing outside our compartment waving a green flag and commanded him, 'Stop the train. Please.'

The Negro guard brought the green flag down, pulled out the red flag from under his armpit, waved it in the air, and came towards us. The creaking sound that the wheels had made as the train started abruptly came to a halt, and an argument started between the white and the Asian woman. The Negro boy cringed helplessly in his corner; even the railway guard seemed helpless.

The Asian woman explained, 'I'm ill, and I need this servant to look after my children. That's why I've got him a reservation in this compartment.'

'No, he cannot travel in this compartment,' the white woman said as if she were ordering one of her servants.

'Why not?' the Asian woman finally asked in the same tone.

'Because ... because ...' she hesitated and said, 'because the servants travel in a separate class.'

I knew she wanted to say something else—'because he is a Negro'—but living in the land of the Negroes, she couldn't well have said it.

The Asian woman said, 'Well, madam, if this bothers you so much, you can change your compartment. And, in any case, he is not taking your space.'

The white woman looked towards Mommy, as if expecting her to say something on her behalf, but at that very moment, Papa started playing with the youngest child of the Asian lady, and Mommy took out bars of chocolate from her purse and started distributing them to the children and to the Negro boy.

I knew this was Mommy's silent response to the white woman.

The white woman looked like she was about to have a fit of weeping.

Her two companions who were still waiting on the platform looked at us with a scowl. One of them started helping her down from the train, and the other picked up her luggage. I heard one of them say to her, 'You can go by air, *mi'daire.*'

But I'm sure the Negro guard found her room in some other compartment because, as the train left, I saw the two men leave the platform without her.

After this little incident our journey was pleasant, and we came to this town which does not have electricity or tap water. But Mommy, Papa, I, Tina, Fiona—each of us likes this place because we can move around alone wherever we want to and because everyone here knows everyone else—the Negroes, the Asians, the white people, and even those whose skin is yellow like the underbelly of a gecko, but who have short crinkly hair.

I've heard people say that the mothers of these people were Negroes, and their fathers Europeans who, when the British rule came to an end, went back to Europe.

How can that be possible? This, too, was one of those things that stayed in my mind and bothered me. How can a father leave his children and go away forever? Once when I asked Tina this question, she answered as grown-ups often do: 'Strange, isn't it?'

I knew that she did not want to answer my question, or that she did not herself know the answer. She would argue with me on such matters only when she had something to say to me to shut me up, something she had perhaps read in a book some days ago. At such times, her manner of speech would always become that of the grown-ups.

One day when Papa was away on a tour and Mommy was busy looking after her patients at home, Tina came to me and whispered, 'Do you want to see someone being buried?'

'Who died?' I asked impatiently. 'Let's go.'

She put her finger on her lips to keep me quiet and answered, 'A driver from Papa's team.'

'Which one?' I asked eagerly.

'You'll see soon enough,' she said and motioned me to follow her.

Tina had no school that day. Fiona was outside playing with the dog, and we were afraid the two of them might want to accompany us if they smelled us. But Tina had the whole trip carefully planned in her mind. At that time she looked more like one of the characters from her adventure stories than the sister I knew.

We left the main road and walked through a corn field, and entered the hospital, which was about a mile away from our house, by climbing the boundary wall.

At a little distance from the hospital was a deserted building whose door had a lock on it. She took me to that building. Standing on her toes, she peered into each of the windows and then asked me to look inside one window.

Inside, on a white tiled table, lay one of Papa's drivers, the one who had brought me a pet monkey a few days ago.

There was blood on the table. The top of the man's head seemed to have been cut open and then stitched with thread. Tina said that that morning when his two wives went to wake him up, he was lying dead in bed, and Mommy had said that an autopsy was to be done on him before he could be buried.

I asked her what 'autopsy' meant, even though in doing so I had to hide my humiliation.

'It means to find out if he was poisoned and that's why he never woke up from his sleep. But Mommy says he had a heart attack. She found that out after he was cut open. That's autopsy. His body has been sewed back again.'

I moved away from the window. I didn't want Tina to know that I was feeling scared. His burial was some time away, so we came back home.

An hour or so later Tina came to me again and said, 'Come on.'

Going through the corn field, we walked towards the Muslim graveyard.

We could see the whole scene from behind the thick shrubs. I saw Papa standing, looking downcast, among the people who were gathered there, and, in a way, I felt relieved that our stealthy trip would not be discovered, for we had left Mommy and Fiona asleep at home, and Papa was standing there, right in front of us.

The driver's body, wrapped in a white cloth, lay on a wooden plank. Blood was still oozing out of the white cloth.

Two men nearby were digging a pit. The red soil which they had shoveled out lay in two mounds on both sides.

Then some of the people filed together in a straight line and went through what Tina later told me was the final service for the man, whatever that meant.

Papa and some men stayed on one side while this service was going on. I knew the reason for this, but Tina also told me. Papa was a Christian, while the driver and those who had come to bury him were Muslims.

Some chameleons moved about in the bush around us, but I was more afraid of the red ants which were close by.

A little later, the people picked up the driver's body wrapped in the cloth and placed it inside the pit. Then they spread some banana leaves on the body and began shoveling in the soil that had been taken out earlier. Soon the two mounds disappeared. In their place now there was a long grave around which the people stood. Everyone, including Papa, raised his hands to say a prayer.

Tina once again looked at me the way she does when she wants to show that she alone knows all the secrets of the world, and whispered in my ear, 'Six feet. See? From his head to his toes, that was all the land he needed.'

I was grateful to Tina for having brought me there to watch the burial, but even then I hated her sermonizing.

In the evening, in Tina's presence, I asked Mommy, 'Was Tolstoy really a great writer, Mommy?'

Tina's ears turned crimson.

Mommy said, 'Yes. He was truly great.'

'Even greater than Johanna Spyri and Enid Blyton?'

Mommy seriously said, 'Maybe not yet, but when you grow up, perhaps he may also grow with you to become a greater writer than Johanna Spyri and Enid Blyton.'

'And whatever he writes, is it absolutely true?' I asked.

'Yes.'

'Is it, then, true that a man needs only six feet of land?'

'In a way, yes, it's true.'

I think Tina had started sweating by this time, but she did not move from her seat.

'Then, why...?' I stopped.

'Why what?' Mommy asked in an exploring tone.

'Oh, nothing,' I tried to evade the subject.

'It seems there *is* something,' Papa joined in the conversation for the first time.

He had been quiet since the evening. The driver's sudden death had obviously touched him. The driver and Papa had been together on many long trips through the dense forest which is dark even in daytime, and where one is afraid of attacks by wild animals and by the enemy tribesmen who might recognize the driver by the marks on his face and try to kill him. In such situations, Papa would take over the wheel and the driver would hide between the seats in the back. At eating time, if the driver's food was finished, he would start eating from Papa's plate—if Papa wasn't eating pig's meat—and sometimes even Papa shared the driver's food.

I knew that, like me, Papa was also feeling sad because, in the afternoon, the two of us had been patting the back of the monkey the driver had given me.

Then I almost exploded, 'It's all lies, all lies.'

Mommy and Papa looked at each other.

'What's all lies?' Mommy asked me. 'Do you mean all that Tolstoy has written?' she asked again, politely.

'Has anything happened?' Mommy then asked Tina.

Tina bowed her head.

'Okay, okay. We'll talk about it after supper,' Papa said.

I wasn't sure if I was sad and crying or just angry. Also, if I was angry, was it at Tolstoy or at Tina? And if I was sad and crying, was it at the driver's death or some other thing?

Images of people—my grandmother, the white woman whom we had met in the train for a while, and dozens of others—were floating in my mind.

I recalled the words of the Sunday morning sermons in our Church and all those things I had heard in the car during our night-long travel last summer from Pretoria to Durban, when we last met grandmother. At our departure she had said to Papa, 'Hank, you are welcome to come back here anytime you want. This is the only country in the world where the settlers had only one ideal—that when they sit outside on the steps of their houses, they should not see the smoke rising from the chimney of their nearest neighbor's house.'

During that journey a friend of Papa's was driving the car. Papa sat in the front seat, and Mommy, I, Fiona, and Tina were in the back. At night time, on a long stretch of road which went smoothly for miles and miles, who could have stayed awake for long?

The hills which during the day seemed wrapped in the green velvety grass were hidden by the darkness. Once in a while the headlights would show where one or another began to rise from the roadside.

We were passing through familiar cities and towns—Colenso, Ladysmith, Pietermeritzburg, all of which were quiet.

The houses of the white Afrikaner farmers stood quietly on the farms, miles away from each other. In one corner of each kraal, very close together, were the quarters of each farmer's servants.

Even during the day I had found the streets of Pietermeritzburg deserted. At night they seemed even more frightening, even though they were lit. They seemed to have been made empty of people.

After talking for a while, Mommy had gone to sleep. Fiona had been asleep since the journey began, and Tina couldn't keep awake for long, for she couldn't read.

Papa and his friend were the only ones talking. They were speaking in low tones, saying things which were altogether new for me. For instance, they said that when the settlements in the South had just begun, some woman had written in her journal that the policy of equality for blacks and whites was against the teachings of the Bible.

What 'policy' meant I did not know, but I knew that what the Sunday sermons in our town said was different from what that woman had written.

Another thing they talked about was the belief among the white people in the South that it was against God's laws to give equal rights to blacks and whites, and that it was every free man's birthright to acquire as much land as he wanted.

Much of what they said at that time I couldn't understand fully, but, as always, I enjoyed listening to Papa. Even his friend talked like him.

They talked about the original people of Australia, about the American Indians, about the Israelis who had come from Europe and about the Jews who were non-Europeans.

I felt as if all the countries of the world had been taken over by the white people, each one of whom was running, like Pakhom, to possess as much land as he could, even if in the effort he had to destroy the Negroes, the American Indians, and many other dark races of the world.

I woke up when the car stopped at Durban. In the early light of the day I saw Papa's friend shaking hands with him and with Mommy.

Before leaving, he said, 'So, Hank, not coming back to this gorgeously beautiful suffocating country?'

Papa said, 'No, thanks. But you look after yourself.'

Some time later we came to know that that friend of Papa's had been sent to prison.

After supper Fiona was sent upstairs, and Papa and Mommy listened to Tina tell them about her little adventure of the day, the adventure in which she had been the leading figure. By that time I had overcome my anger or whatever it was that had made me feel so miserable earlier. I was in control of myself.

'So, what is it, son?' Mommy asked me.

'What?' I asked.

'What has your adventure of the day got to do with Tolstoy?'

Hesitatingly, I repeated the things I had heard during our car journey from Pretoria. Mommy and Papa watched me in amazement. I think they felt proud of my good memory.

Then, suddenly, I asked them the question that I had been meaning to ask since that afternoon, 'Maybe, as you say, Tolstoy has written good stories, but I think Tina lies when she says that a man, from his head to his toes, really needs only six feet of land. My question is: How much land does a white man need?'

—*Translated by Faruq Hassan*

Emancipation

If there is anything in life worth hating, it is hatred. And that's been true with me—always. Long ago, my father and my mother often flew into a rage over my failure to hate certain people; later, my in-laws.

My husband and I were returning from a ritual bathing in the Ganges. My in-laws knew why we had gone there. They knew that I was barren and they hoped that a few drops of the sacred water might help me get pregnant. But my husband and I knew well that the visit could not help us. My husband cared a bit too much about religion, otherwise he would not have set out at the mere suggestion of others on a costly pilgrimage to the Holy Ganges; he was far too shrewd to dump money into a worthless enterprise. Perhaps he believed in miracles.

As for me, I've always loved the Ganges. I still do, with reverence. In fact I love all rivers, not because loving rivers is in my blood, but because rivers have informed my childhood.

Travelling by train between east and west we passed over the Ganges many times. Each time, just before the arrival of that moment, my mother would sit up tense with anticipation, her fist full of coins. Each of us also held some change in our hands. The moment the bank of the Ganges appeared in view, the women who sat in the middle row in the compartment would move over to window seats. My mother had never been negligent in this matter. She always took a window seat, even if that meant sitting out and waiting at the station for all of us under the open sky till midnight.

I enjoyed every part of this journey and felt particularly joyful when the train crossed a bridge. But the joy of crossing the Ganges was in a class by itself. The orange iron girders suddenly rose to high heaven and, just as swiftly, swooped back down, reverberating with strange noises. Just then I would think of all the kids in our city who had never seen a bridge sway in this manner. They could not have been more unfortunate!

Through this echoing rumble would rise the piercing whistle of the locomotive, as if paying its respects to the Ganges, like the women in our compartment. Precisely then the coins would hit the iron girders with a clink and then plop down in the waters below. People in other compartments would also toss fistfuls of coins. I would poke my head out of the window and look back at the long line of compartments behind ours, straining to figure out which among the outstretched hands belonged to my father. Once, though, when a coin ricocheted off a girder and hit me on the head, I jerked my head back inside. I never again stuck my head out. If we happened to be crossing the Ganges in the daytime, I would content myself with peering at the boys who lurked, half-submerged, in the water below the bridge, ready to dive and retrieve a falling coin.

I knew my younger brother always managed to wedge a coin or two between his fingers so that it wouldn't fall. This was our mutual sin against the Sacred Ganges, mine in that although I knew about his deception, I chose not to snitch on him. In principle, I should have hated him for it, but I could not. I was powerless.

Often when the train crossed the bridge at night and I slept, mother would shake me violently to wake me up. The river always looked different at night. The iron girders raised and lowered their heads as always and the river below appeared perfectly calm. Every once in a while, though, the edge of a wave gleamed momentarily in the light of the moon; sometimes the moon itself seemed to have descended into the waters; and often the light from a bonfire filtered through the groves on the river bank. I would let my head, heavy with sleep, rest on the window sill, dimly aware of the clinking coins as they struck the girders and dropped into the water.

But that was a long, long time ago. On the night I am talking about, I sat alone in the compartment. We were returning from a pilgrimage to the Holy Ganges. My husband—my Lord Husband—was in one of the compartments ahead of mine. He disliked the idea of women travelling together with their husbands in the same compartment, otherwise he would not have abandoned me all alone.

After bathing in the Ganges I was feeling a gentle warmth throughout my body; the cool night air blew over me and made me dizzy with sweet inebriation. We were four or five small stations away from our town and still had an hour or two to go. I was in no particular hurry to get home. I had no children waiting for me; as for my in-laws, they were waiting less for me than for the vessel in which I carried the sacred Ganges water and which was now my sole prized possession. But he who was bound to me by the sacrament of marriage, he, I knew, would treat me like a stranger as soon as we reached home. He had repeatedly told me that too much closeness with his wife interfered with a man's contemplative life. He certainly had reached the age when contemplation behooved a man.

I do not hate life. Yet, in spite of my youth, I had become thoroughly bored with its unmitigated monotony. At home, there was nothing enjoyable for me to do, nor were there any books I would have liked to read. You don't expect to recite the *Bhagwad Gita* day in and day out.

At some small station the train came to a jolting halt. My eyes fell on the ticket checker outside. Outfitted in his black uniform, he stood leaning against the lamppost, his eyes riveted on me. An old-style kerosene lamp, with the name of the station inscribed in three languages on its square shade, burnt palely on top of a post. In its dim light, I could scarcely see the ticket checker's face. The train made a scheduled stop at this station at night out of sheer formality, for hardly anyone boarded the train here even in the day time.

My husband wouldn't be impatient to check up on me. We had been married for three years, and he had long ago passed the stage when he could be expected to come rushing to inquire about his wife every time the train stopped.

I glanced at the ticket checker, then at the vendor, and then at the dog sleeping under the vending cart. After I had checked the name of the station, my eyes grew heavy with sleep. My body was still warm after bathing in the river and the night air was pleasantly cool.

Dimly through my drugged senses I heard the guard whistle, the engine pump out a few quick jets of steam with a piercing squeal, and the ties hitching the carriages together groan as the train moved out at a snail's pace. Then I must have dozed off. When I woke up with a start, the train had moved quite far from the station and picked up speed, and the ticket checker was bolting the door from inside, his back to me. I don't think of myself as an atheist, still I didn't think it necessary to invoke the help of the holy water I had with me.

That snatch of sleep must have been the longest sleep I ever had. Just as the end of sleep signals the end of night and the arrival of a new day, so on one end of that sleep lies the dark night of my life's story and on the other, its new dawn.

Instead of sitting up in alarm, I just lay there and through my barely closed eyes watched him approach me almost without a sound. He was in his mid-thirties, of stout build, with a fair complexion—perhaps, for that reason, quite handsome by some standards—probably a native of the northwestern region. When he had come quite close to me, I pulled myself up suddenly and asked him boldly, 'What?' To appear undaunted by danger, even if there were any, I gazed out the window unperturbed.

Until a woman looks directly into a man's eyes, he doesn't know quite how to use force, like someone poised to attack but unable to find the pretext to do so. My nonchalance disarmed him so completely that he asked nervously, 'Ticket!'

I knew my husband had our tickets, even so I started fumbling in my purse absent-mindedly. And then I looked at him with questioning eyes.

'Never mind,' he said, forcing his eyes into those of mine, as he grabbed the upper berth with both his hands and bent over me, blocking all possible avenues of escape.

Hatred for my husband washed over me like a tidal wave. Abandoning me to my fate, he must now be sitting in the men's compartment without a worry, supremely satisfied with his great piety. He had no right to expect anything from me now—absolutely none.

But women do not lay waste to their homes out of hatred for their husbands.

The vessel of the sacred water lay some distance from me. I looked at the emergency chain helplessly. There was no way I could reach it, at least not without a scuffle. And there he was, looking at me in a manner calculated to captivate me with his strong, masculine beauty.

Suddenly, as I sat surrounded by him, a scene from childhood came rushing back to mind. I was travelling by train with my mother. At some time in the night, the train pulled into a small station somewhere in the east. A haggard, breathless peasant forced his way into the jam-packed women's compartment to help his wife disembark. The couple gathered their baggage—a few large bundles and some tin canisters—and began offloading it from the carriage. The husband grabbed their small boy, climbed down, sat the boy on top of a bundle, and climbed back up. Just then, the train started to move. We all heard the boy's pathetic screams as he sat in the darkness at a station which didn't even have a platform.

The husband was in a fix: should he jump off the moving train or stay aboard with his wife? Somebody suggested pulling the chain, which they did, repeatedly, first the wife, then the husband. But the train, unaffected by their misery, chugged along. The woman began to wail, the husband comforted her, 'Don't worry. We'll get out at the next station, rent a bullock-cart and go back. Why are you crying? Surely somebody will take care of the boy till we get there.'

I did not resist, which encouraged the ticket checker to sit down beside me. The strong scent of eau de cologne wafted into my nostrils. He seemed to have doused his chest rather generously with it just before entering the carriage.

I felt his hand crawl slowly across my bare back and fall on my other shoulder. The speed of the train had not dropped, nor could the familiar sound of changing tracks, which usually announced the arrival of a station, be heard. I felt the weight of his body over mine. Through my fogged senses I managed, God knows how, to say, 'But I am not clean!'

At that level of intimacy force was altogether unnecessary, and so was violence. He laughed and got up to go to the opposite seat. In that split second, when his back was turned to me, I made a dash for the chain and pulled violently at it, just as the peasant woman had done. The ticket checker's last, unfinished sentence, 'Where do you ...' cut off in a hurried, nervous 'Stop! Don't do that!'

He jerked at me, trying to pull me away from the chain.

The sound of the train coming to a sudden, grinding halt rose in the deathly, dark jungle—a sound evocative of the clanking of chains as they were shaken off the body of some prisoner. The ticket checker threw himself on my feet.

I was crying. My hands were still clutched to the chain. I heard voices approaching us outside in the darkness. He entreated, 'Please forgive me. Please don't tell them anything. For God's sake, please ...'

At that moment he who was so strong, tall, fair and— perhaps—quite handsome, looked utterly miserable, so pitiable in his venial helplessness lying on the floor; even more pitiable than I had felt myself to be just a short while ago.

'For God's sake ...'

For a moment I felt the same hatred writhe inside me, which my educated parents had always encouraged me to have, and which my Lord Husband and his parents had always considered a part of religion.

Then the sound of the door being forced from outside was heard. The inside door handle turned, as though by itself. The ticket checker got up, slowly dusting off his uniform. The struggle had ended for him.

I peered into the darkness outside. I knew this area. Further down where a feeble light filtered through thatched huts, was the place where I had often seen peacocks pecking at grain during the day. Give or take half a mile, but no more. But why was I thinking about that? Really, such a thought could not have been more out of place!

The railway guard and police escorted the ticket checker out of my compartment. Both of us were doing our best not to look at each other. The sound of heavy boots crunching the gravel by the tracks could be heard for some time, and then the train started again.

I was ready to do anything, just anything; but what I could not do was to cry on my husband's shoulder. His shoulder would have been a lifeless object for me to rest my head on and cry. Sitting opposite me he may have looked like religion incarnate, but to me he was no more than the god of hatred; a man who could hate others, singly or collectively, by calling them Muslims, because they belonged to this country or that, because he assigned them to a despicable caste; and who could also hate a lonely and vulnerable person like myself because I was incapable of producing in my stony heart an emotion as delicate as hate.

What could he be thinking, I wondered. Being disgraced? But if he was disgraced I was hardly to be blamed for it. A certain satisfaction, because my assailant was, as my husband had suspected all along, a man of another faith? Or, perhaps, brooding over the poison which would continue to spread inside him; what if that abominable barbarian has succeeded in what he had set out to do?

When I next saw the ticket checker he was handcuffed. He was under arrest at our station. A sudden desire to go up to him and smell his chest overwhelmed me. But I restrained myself, fearing I might break into laughter. Instead, I buried my face in the handkerchief, as one about to cry does.

After a brief interrogation the case was entered and we were allowed to go home.

That pious visit to the Holy Ganges changed the course of my life.

My husband assumed an attitude of chilling aloofness and silence. My mother-in-law treated me like a Shudra. Whereas earlier I could denounce certain Hindu customs with impunity, I could no more. Even if I had criticized the outmoded practice of female self-immolation, it would now have been interpreted as an ill effect of my Western education. I even avoided going anywhere near the home altar which had the image of one of the gods. For one night, after a protracted silence, a question echoed in the darkness: 'Did he touch you?' And then I heard a long gasping sound. They were words that betrayed a long festering doubt, a doubt that could only now accept some reassurance.

I had already been subjected to the interrogations of the judge and the officers investigating the case, to which was now added this accusation, rising through dark silence. Certainly this was not my own conscience reproaching me; that poor voice was silent. There was nobody with me in the room except my husband; and my conscience slept peacefully like an innocent child. My husband lay so far away from me that even if we had extended our hands we could not have possibly touched each other. Suddenly, I was overcome by the desire to break into laughter and ask: 'That's *it* then. Your Lordship has been agonizing all this time over whether someone has touched me with the same intention as you once had yourself?'

Even now when my eyes fall upon an ad for used cars and the words 'Owner Driven' or 'One Man Driven' come to view, I am suddenly reminded of someone exhaling a long, gasping breath in the darkness and of my totally unnecessary reply, 'No. No one has ever touched me except you. I am unblemished.'

A little while later I heard my husband turn over in bed and snore. His snores sounded like *'Hare Om.'*

One winter day we were sitting in the sun in the open veranda. I was knitting a sweater. My mother-in-law was rubbing oil into my husband's hair. My father-in-law was ensconced in his rocking chair reading the newspaper. The next day I was scheduled to appear in court. At every court hearing my in-laws, all of them, felt they were being disgraced somehow; as their daughter-in-law would now have to get into a tonga and go to court where she would be asked about criminal assault, rape, and other indelicate matters. And as I was subjected to these questions, my husband and father-in-law would sit through them with their heads bowed in shame. They would try to stay the farthest away from me, so that the world would know that the law of their religion was infinitely more important to them than the law of the court. No matter how the court ruled, as far as they were concerned I was no more than a beautiful, expensive glass object, which, once broken, is allowed to remain at home, but which can scarcely be used again.

To break the silence, my mother-in-law asked my husband, 'You never did finish telling me about that yesterday.'

'About what?'

'You know, you were telling me about your friend who was forced to go out of business—remember?'

'Oh, well, he had a quarrel with his partner.'

'I already know that. But when your friend was setting up the shop he found only mountains of good in his partner, didn't he?'

My husband looked at me, as though I were expected to know the answer.

'No good can ever come of these people,' my mother-in-law observed. 'Never!'

By now my father-in-law had stopped rocking in his chair. He lowered the newspaper and looked at me over the top of it with questioning eyes.

I could not have been more absorbed in my knitting than I appeared. But finding everyone staring at me, I scratched my temple with the knitting needle and tried to say what they wanted to hear from me, 'Your friend should not have gone into partnership with a Muslim in the first place. That was a mistake.'

The colour of suspicion turned a shade darker in their eyes.

I continued, 'Muslims simply cannot be trusted. Never!'

I heard the echo of my thoughts: You are telling a lie.

My father-in-law quietly got up from his chair, left the newspaper on it, and stepped into the backyard garden. My mother-in-law, remembering some unfinished work, hastened to the kitchen; and my husband settled down into his father's empty chair and began browsing through the newspaper as he rocked gently.

I knew they thought I was lying. Yet, in those days, I found myself saying only those things that pleased them: how ridiculous the customs of Muslims were, how deplorably unclean they were, how at one of my Muslim girlfriends' house everyone walked right into the kitchen with shoes on. Like everyone else at my in-laws', even I started calling the despotic father-in-law of my husband's sister 'Aurangzeb', the most fanatic Muslim ruler, because he was an utter kill-joy, who enjoyed hurting people indiscriminately and went after them with a vengeance.

But when my sister-in-law showed up one day at our house and I asked her half in jest, 'How's Aurangzeb-ji?' nobody found it funny any more. Perhaps the joke had gone stale, or had become out of place.

My Muslim girlfriend still came to see me occasionally. On such visits, I tried not to talk with her alone. Instead, I would make her sit within earshot of my mother-in-law. If, during our conversation, she slipped in an English word, I made sure that my next sentence included its meaning in Hindi. If she asked in English, 'How is it going?' I would reply in Hindi, 'Who knows how many more times will I have to appear before the court yet.'

After talking in Urdu for a while, if she inadvertently used an English word, say, 'witchcraft,' I would say, 'I don't believe in this *jadu-tona* stuff. Maybe they do.'

'They' stood for the Muslims—of course. At the back of such unnecessary clarifications loomed, perhaps, my fear that in spite of the calamity, by letting me continue meeting with this Muslim friend, deep down in their hearts my in-laws had consigned me to the category of the irreligious or as someone contemplating conversion to Islam; or if not that, then surely as ready to run away.

But on such occasions my mother-in-law invariably managed to leave the room. My husband, if accidentally he came into the room, would ask an unnecessary question—such as where he could find a particular book—and then exit right away. He hated us both.

One day I told my Muslim girlfriend, 'There are always two elderly men with my assailant at court hearings. Both give me the creeps.'

'Must be Gog and Magog.'

'One of them resembles my assailant quite a bit. He is the same height, the same build, the same fair complexion. The only difference is that he has a beard.'

'Must be your assailant's father.'

'And the other man, who is even more suspicious, always waiting for an opportunity to stare me in the eyes, he also has a beard—who could he be?'

'Your assailant's father-in-law. Who else?'

'Can't be,' I said, categorically. 'I think I have seen his father-in-law. He must be the same man who was trying to comfort my assailant's wife by calling her, "Daughter! Daughter!" That was when she saw her husband in handcuffs for the first time and started crying. Some woman! I don't know if I could have cried had I been in her place.'

'She must be crazy.'

'Strange thing, though, is that she was looking at me accusingly instead, not at her husband; as though I was the one who should have been handcuffed, not her husband.'

'A husband is always innocent in the class she comes from. Surely you know that.'

Besides a few custodians of the law, there were just the five of us in the courtroom: my husband, me, the ticket checker, his young wife, and the wife's father.

Even on such a shameful day the three stuck together. Perhaps they were thinking that I was solely to blame for this misfortune. Perhaps the wife thought I had falsely accused her husband, or even if he had in fact assaulted me with criminal intent, so what? The offence was not grave enough to warrant his being imprisoned.

The father did not have the heart to see his daughter in tears, nor the wife to see her husband in such straits. I had the curious feeling that this incident was no more meaningful to them than algae spread over the waters of their life, which they preferred to tear away so as to drink again from the limpid waters beneath.

They wanted to settle with me out of court, were willing to pay the damages if I could be persuaded to say that being all alone in the compartment that night I had panicked at the sight of a man and pulled the chain out of sheer nervousness.

In the eyes of my assailant's wife I saw an earnest entreaty. I also saw scorn and accusation in those eyes, but what I did not find was the slightest trace of the kind of feeling a woman is expected to have for her husband under such circumstances.

To settle out of court, to offer a bribe, to apologize, or, if need be, even to destroy me—they tried every method. At least they were unswervingly united behind a common purpose.

But for our part, it was as if an animal had died in the pool of my life and that of my husband. With every passing day, the dead body—bloated, decaying, deformed—rose to the surface. To drink from such foul water was out of the question.

After the first two court hearings, in which my assailant appeared accompanied only by his father-in-law and his attorney, and which ended in adjournments, suddenly, at the third one, he appeared with a crowd of supporters who had flocked from God knows where. Broad-chested, young and old men, who looked so much alike: matching turbans with black skullcaps poking out of the center; the same silver and gold chains dangling from gold and silver buttons in their lustrous black velvet waistcoats; the same fresh, rosy complexions; and the same baggy *shalwar* trousers flapping in the wind as they walked. Even their sandals looked curiously alike. Only their ages and the presence or absence of beards and moustaches made any difference at all.

Those two elderly men were part of this crowd.

The two would bring things to eat, and would try to feed them to the ticket checker whenever they had a chance. Unlike my assailant's wife, they didn't feel sorry for him, but neither did they seem to feel any apprehension over the bleak future which awaited him. They would offer him advice all the time and steal sideglances at me whenever they could. I could do nothing about it, except cover my head even more securely with the hem of my sari and try to hide myself. As it were, I had to fend for myself in the court; no one was there to sustain me in my ordeal. In a manner of speaking, for me at least, the trial had ended that fateful solitary night after I had pulled the chain. More than this, I neither hoped for, nor cared about.

Before the crowd of my assailant's supporters materialized, I heard a rumour that a telegram had been sent to his village and his relatives were about to arrive. Later on I came to know that one of the two old men was his father and the other some accomplished Muslim holy man. The latter looked fearsome, and always seemed to be mumbling something. He had a ridiculous moustache; it was trimmed so that it began about half an inch above his lip. His beard was red, and so was the hair that streamed down over his neck from underneath his turban, like a horse's mane.

In the courtroom, this old man tried to cripple me with the piercing intensity of his eyes. Once in a while, finding him absorbed in meditation, I would sneak a look at him. The strange thing was that underneath his hairy appearance, he did not differ much from a Hindu sage—the usual *rishi munis* I knew. Both have a kind of tranquil fire in their eyes. He had been especially summoned to help the ticket checker out of the mess he had created for himself. Before answering a question, the ticket checker always looked in his direction for an approving nod.

My Muslim girlfriend later told me that this old holy man was an adept in a special charm, which is read as one slits the throat of a black rooster. Nobody is allowed to eat the rooster, and it has to be buried while still writhing and fluttering. When I heard that, an unknown fear made me involuntarily touch my throat.

'And what does he recite in the charm?' I asked, trying to drown my dread in a laugh.

'Oh, he just asks for succour from Bhagwan, or, let's just say, from your Bhagwan's Muslim counterpart.'

'Does He help?'

'Certainly.'

'Even in a case like this?' I asked. 'I mean even people like my assailant?'

'Surely you don't mean to say that evil people have a separate god to ask for help, do you?' my friend scoffed.

I was living like a perfect stranger at my in-laws'. Once it occurred to me to end the episode where it had begun. Why not another pious visit to the Ganges? But then I thought that for that purpose, the river in our town was just as good. Almost every year it provided peace to some agitated soul or another.

One morning in the month of Baisakh I went to the bridge over our local river. Below, by the bathing ghat at the edge of the water, I could see women's yellow and red saris. Children sitting on the bank were skipping pebbles and rocks in the water. A few boys were bathing directly below the bridge, their black shoulders glistening above the water in the morning sun. With no particular reason I extended my hand over the guard-rail and waved it as though I were throwing something into the water. The boys at once shouted, 'Mataji! Mataji!' and began paddling around in the water.

The memory of my dead mother came to me. I also remembered my brother—the thief!—who was now living in Germany, my sister who was in the U.K., and my father . . .

The breeze over the bridge was cool and comforting. I took all the change out of my purse and threw it into the water coin by coin. With every falling piece the boys plugged their noses and dunked their heads into the water, emerging seconds later with the same coin which they showed to me in their triumphantly waving hands.

I returned home in the afternoon, without having bathed in the river, and went straight to my room.

I don't know who came up with the saw, 'There are as many versions to a story as there are mouths.' But he must surely have come up with it centuries ago in a courtroom. Based upon events and statements piled one on top of the other, the story that finally emerged was far more lengthier and different than the actual experience I had gone through within the space of a few moments that night. When the prosecution went into details of my college days, my religion-preserving husband just sat there dumbly. And his father took a deep breath as though his worst fears were being proven true today.

The ticket checker maintained that: (1) it was true that he was standing against the lamppost, but he was simply filling out the routine papers there; (2) it wasn't true that the lamppost was located directly opposite the women's carriage; (3) when later the train started to move he did see a young woman emerge from one of the carriages toward the tail part of the train, but whether this woman was beautiful he could not have guessed at all; (4) he suddenly felt as though the woman wanted to commit suicide, so when the carriage passed in front of him, he grabbed the handlebar on the door and climbed aboard the footboard, and entered the carriage as he pushed the woman in; (5) the woman was crying inconsolably at that time because, as she said, her husband was thoroughly fed up with her and, along with her in-laws, unhappy over her inability to bear children; and (6) the woman had even told him that she had once before attempted to kill herself, and if, after her present visit to the Ganges, she again failed to conceive, it mattered little whether she lived or died.

Some of these statements were in fact true. For instance, it was true that I was barren. It was also true that during the course of the trial I had often been seen strolling over the bridge, instead of bathing in the waters below it.

But the testimony of the railway guard proved inconclusive. It failed to establish that when the train stopped the ticket checker was struggling with me to keep me from jumping off the train and had managed to pin me down on the floor and clamped both my hands. That Muslim guard had seen me hanging on to the chain, all right!

The case was dismissed. The ticket checker and I were both honourably acquitted and came out of the court. My husband and his father were terribly unhappy; if not for my sake, then, surely, for their religion's sake, they had badly wanted to see the ticket checker sentenced to prison.

I stood outside the court building—abandoned, friendless, unwanted, while my assailant walked away surrounded by a crowd of well-wishers who had joyously garlanded his father and the holy man. For a moment I imagined a scene: some people are digging a pit with worshipful reverence. The ticket checker places a big black rooster in the hands of the holy man, and the former's father, a long knife. The saintly figure mumbles something inaudible. Suddenly I see a jet of blood shoot from the rooster's slit gullet, followed by a long rasping breath. Seconds later a flutter of wings and the black, lustrous wings themselves go down the pit and disappear under the fine, yellow dust crumbling over them. The men slap down the dirt with their hands and then stamp upon it with their feet.

Quite as suddenly, I returned to my senses. This was the exoterica of religion, part of its fascination and appeal.

* * *

The river was muddy. Big round patches of oil floated here and there on the surface. The shadow of the bridge trembled over the water on the downstream side. Once in a while a tugboat passed by, or the loud whistle of a ship was heard. *Ghon-on-on.*

An empty beer bottle came floating in the distance. I picked a coin, took aim, but stopped. The bottle was too far away.

A little while later an empty beer can, thrown off the bridge or from one of the ships, came floating our way. I took aim with a coin and hurled it as I said, 'Amstel.'

'No, Heineken!' she said from behind.

The coin hit the can with a clink. The can swirled and the name bobbed before me for a second. Triumphantly I said, 'See, it's an Amstel!'

Both of us laughed. Sitting on the grassy bank we often played that game, especially when it wasn't misty. And this evening was glorious. On both banks of the river, European, Asian, and African children were running around on the grass and rolling down the slopes, while their mothers sat in the mild sun and read newspapers, or just peered through their binoculars to watch the ships come in or go out. A few infants had fallen asleep in their strollers.

'Another round—shall we?' my West Indian friend asked.

'No, that'll do for today,' I said. 'I've got to cook supper for the children as soon as I get back home. My husband is on call tonight. And since today is my lucky day, I hope the phone won't ring every five minutes for him to rush to the hospital to take care of some drunkard.'

Getting up from the grass, I said, 'Your score is four, and mine seven. The river swallowed five of your coins and two of mine.'

She also got up, dusting grass off her slacks, and said, 'You're a perfect marksman. I just can't believe it.'

'Can't believe what?'

'That you still don't eat meat.'

'That has nothing to do with eating meat,' I replied. 'I have been throwing coins at targets in the river all my life.'

—*Translated by Muhammad Umar Memon*

The One Upstairs

The story goes back to the days when some people spent their Sundays like Fridays, and others their Fridays like Sundays; that is, it goes back to eight or nine years, not further than that.

Around 1:30 on Friday afternoon was an awkward time to visit anyone. But what difference did it make thought Yasmeen. Perhaps her visitor was from out of town. Out-of-towners, when they did come to the city, usually had a number of chores to do; for example, they might have a doctor's appointment, they might have to buy new eyeglasses, might have to sell a few things, or make a deal for some big items, might have to talk to a lawyer, to deliver a sack of vegetables to someone's house, to inquire about the health of a relative (their own or somebody else's) in the hospital, to go see a movie, to buy a earring or a necklace for the wife, or knick-knacks for the children, and so on.

And when they had nothing else to do, they made a tour of the area behind the Memon district, among the fruit vendors' trolleys and the flower sellers' stalls, even if that tour was made just out of habit, and was, in reality, without any point or purpose.

So, Yasmeen got up and opened the door. She recognized the man, even though her eyes were only half open because of the glare of the sun. Her visitor was the seed salesman, and had been to see her many times before.

She rubbed her eyes and said, 'At this time?' She abruptly asked, 'What's that under your arm?'

The man standing outside the door was drenched in sweat. Even then he moved his hand gently on the head of the thing he held under his arm and answered, 'Don't you recognize it? You eat one almost daily.'

Yasmeen took a step back and said, 'A rooster?'

'Yes,' he answered, and without seeking her permission walked into the room. 'This is a rooster, and I am Rafeeq Khan. Recognize us now?'

Yasmeen was gradually getting over her lassitude and drowsiness. Walking towards the earthen water pot in a corner of the room she asked, 'What time is it now?'

'One-thirty, or maybe a little later,' Rafeeq Khan said, laughing a little shamefacedly.

She drank some water from a glass, and adding some more to what remained, she handed it over to Rafeeq Khan. She lifted the drape at the window and began to peer out back into the gully, where there was a mosque. She said, 'People are going to say their prayers.'

'I've already gone. There are dozens of mosques. This is the only place around here where you can hear the call to prayer, though, isn't it?'

Then he came and sat on the bed near Yasmeen and, laughing a little more shamefacedly, said, 'I had to come and see you, so I got mine out of the way.'

He started undoing the string around the rooster's legs. Yasmeen sat on the bed leaning against the wall. She had pulled her legs up and encircled them with her arms. Everything that was going on before her eyes was altogether new for her.

Rafeeq Khan took off the string around the rooster's legs and lovingly rubbed them with his hands, as a prisoner's companions rub his wrists when his handcuffs are removed; he tickled the rooster under its neck, patted it on the head, the back, the feathers, and then, gently tying one end of the same string to the rooster's leg, attached the string to Yasmeen's bedpost.

As soon as the rooster could move about freely, it flapped its wings and crowed.

'Look at it! It thinks it's morning,' Rafeeq Khan said.

'Or maybe it wants something else,' Yasmeen said.

'What?' Rafeeq Khan asked, surprised. Then getting the drift of her comment, he laughed and said, 'Do you have a bowl or something? This silly bird is making a call to prayer at its own death.'

'Do they do that?' Yasmeen asked incredulously.

'I don't know. Maybe. My job is only to sell seeds and to come see you when I can.'

Yasmeen got up from the bed and after some effort found a handleless cup. She poured some water into it, placed the cup in front of the rooster, and said, 'Are you going to slaughter it?'

'The one I've brought it for will do that; doesn't look like it's a fighter.'

'Who did you bring it for?' Yasmeen asked.

'Well, first I thought I'd give it to the doctor. He's treated my sister-in-law with so much care—must have given her at least ten bottles of glucose. I was going to see him for the last time today to tell him how she's feeling, and I thought I'd take this for him. But it doesn't look like it's meant for him.'

'Why?'

'The clinic is closed; he's gone somewhere. Maybe his own brother-in-law has died,' he said and laughed.

Yasmeen sat near the rooster and began rubbing her fingers on its head. Earlier, at the door, her reaction upon seeing it was not one of surprise, but of fear. In her adolescence, which was not too long ago, only six or seven years back, she had often raised chicks, red and yellow ones, each one bought for a rupee. She had cried when, one after another, they died, for there was nothing else in the house to amuse her. People came to the house, sat with her mother or sister, talked to them in different languages; she knew, as a trained nurse does without being told what instrument or needle or swab the surgeon needs, when to leave their company, and would quietly slip away and go upstairs.

Some of those who visited her house were such bastards— this word was used in her house as commonly as words like 'greedy' or 'generous' or 'true to one's word'—that despite having beards which looked like mottled grass, they wouldn't hesitate to grab and paw at her, as though estimating her fleshiness.

Rafeeq Khan lifted the side of his shirt, took out a wad of bank notes from his waist pocket and examined them. Some notes were loose, separate from the wad. He passed one of these on to Yasmeen.

Yasmeen pointed towards her pillow and went on stroking the rooster lovingly.

Rafeeq Khan placed the bank note under the pillow. Then he lay down on the cot thinking. His hand accidentally went under the pillow where there was some change and a watch.

He brought the watch out from under, looked at it, and said in astonishment, 'Oh God! It's four o'clock!'

Yasmeen burst out laughing and said, 'This watch is fake, only for decoration.'

'Really,' he said and pressed it close to his ear. Then, playing with the watch and counting the roof-beams, he went to sleep.

'Don't you want to give it something to eat?' Yasmeen asked, but his snoring, instead of stopping, became louder, as though after being dead tired, he had come to a very peaceful place where all his cares and responsibilities had ceased, and he had effortlessly gone to sleep.

Yasmeen was now getting tired of sitting on her haunches. There wasn't even a decent chair in the room, so willy-nilly she sat on the foot of the bed, her head resting against the side wall and her legs pulled up to her chest and circled by her arms.

Soon, she began to nod off.

Her visitor's watch lay on the floor near the rooster and his briefcase also stood nearby. The deep henna-dyed rooster, whose feathers at places showed a dark green or blue fringe, had walked around the room as far as its string would allow, and not finding anything worth eating, was now standing on top of the watch. The sole of one of Yasmeen's feet was near its head; it tickled her foot with its beak.

In response, she lazily leaned her back against her visitor's hips. From outside, somewhere very far away, at intervals, the call of a *falsa*-vendor came into the room. But neither that call nor the occasional sputtering of a scooter's engine in the yard could disturb their sleep.

The rooster jumped up to sit on the briefcase, but found it difficult to keep its balance; so, flapping its wings, with a loud bang it came tumbling down to the floor along with the briefcase. Its sudden shrieks said for sure that one of its legs or wings had been squashed under the briefcase.

Startled by the sound, Rafeeq Khan woke up. So did Yasmeen. Both were drenched in sweat, as often happens during siestas on summer afternoons.

A little later, after Rafeeq Khan had offered the rooster some sample seeds from his briefcase and Yasmeen had refilled its upturned cup with water, she asked, 'So what happens to it now? The doctor isn't at the clinic; will you take it to his house?'

'I don't know where his house is. But it doesn't matter. If I don't offer it at one shrine, I'll offer it at another. They're all the same.'

'One shrine is the doctor's; whose is the other?'

'The lawyer's. The bugger has left my case hanging for three years.'

'So you'll go to his shrine now?'

'Yeah.'

'Where does he live?'

'Why, are you in some trouble? Come on, tell me, I can get you help.'

Yasmeen kept quiet.

'The police don't bother you, do they? Or the pimps? Not in a wrangle with someone, are you? But you don't look like that kind of a girl. I knew that the first day I came to you, five Fridays ago. I said to myself, this kid is different. She can't even tell the good from the bad. That day you gave me something to eat and you also massaged my feet. And when I was about to leave, you remember, I stood and waited in the doorway? Maybe you don't remember. I don't blame you. Do I remember all my clients? I thought maybe you'd ask for some money for milk, that's why I'd stopped in the doorway, but you just went to your bed and lay down there. You looked a little flushed that day.'

Yasmeen kept watching him with a look of gratitude.

Then with an effort she put her hand on Rafeeq Khan's hand.

Rafeeq Khan placed his other hand on top of hers and kept looking at her lovingly. There was neither any craving in his heart nor guilt in his eyes; in fact, even the little regret that he had felt in the beginning, about having come, of all places, to Yasmeen's, on a Friday, was all but gone now.

It was peaceful all around—no noise of the motorbikes, nor of the haggling clients in the street below. Unlike some young men, or even middle-aged ones, who go to railway platforms to feast their eyes on beautiful faces and then, satisfied with a little ogling, go their own way, the clients in the street below were mostly frightened creatures, often out of breath and panting, their eyes open but their blinkers down, so to speak, for fear someone might be watching them, disappearing in a flash, briskly, either inside to the world of pleasure, or outside, well away from it.

The only sound audible was that of the rooster pecking.

Then Rafeeq Khan put his hand affectionately on Yasmeen's shoulder and said, 'Today I want something different.'

'What?'

'I want you to cook this rooster. Let you and I both eat it together.'

Yasmeen was wide-eyed with surprise.

'Want me to cook it?'

'Yes, I do.'

'And you want to eat with me...?'

'Yes, of course. Why, is it forbidden to eat your cooking?'

Yasmeen's face fell. Rafeeq Khan felt embarrassed. He said, 'Oh, oh, what did I say there? I meant just as all the other women of a household cook and everybody eats, in the same way, I thought ... Did I offend you? I didn't mean it; it just came out of my mouth. I don't see any difference between you and the other ...'

Yasmeen waited to hear him say more. But just as earlier he had said more than was necessary, now he was unable to say what needed to be said. He laughed and said, 'So, what do you say? I can slaughter it for you if you want to cook it. It's a real home-grown bird. My kid cried when he saw me carry it—

where are you taking it, he asked. Even my wife said if you had to please the doctor, buy him a farm-grown chicken from the store. Have we not been paying him the fees, she said? But I said, the doctor couldn't have tasted a *desi*, home-grown one. But today, you eat it. I'll do something for the lawyer later on. Every week, either my father-in-law or one of my cousins takes a basket full of onions or peppers or okra to his house. If he's lucky, he'll get a *desi* chicken as well. But today, it's your turn to eat, and to feed me also.'

Gently, Yasmeen asked, 'How many times have you been here, Khan Sahib?'

'Been coming here for five Fridays now, with the exception of one. You add it all up.'

'And since when have you been coming to this district?'

'I don't remember now. Since before my wedding.'

'Did you ever see a stove in anyone's house? I mean in the house of someone like me?'

Rafeeq Khan pulled his hand away from Yasmeen's shoulder as if someone had splashed acid on it.

He thought for a while and answered, 'I haven't noticed. There have been stoves. But if you say so, maybe not. I don't go around peeking into the different corners of people's houses.'

'There aren't too many corners to peek into,' Yasmeen answered, a little curtly. 'Both our meals come from the restaurant. In the morning we get a cup of tea and a bun.'

'And what do you get in your meals?'

'Oh, what we've been allowed. Meat and potatoes, lentils, oven-baked flat-bread.'

'Meat and potatoes all year round?'

'No,' Yasmeen smiled and said, 'I used that as an example. Actually, we eat what the madam upstairs gives us.'

Rafeeq Khan laughed and said, 'For everybody else, it's the God above, in the sky, who feeds them. For you it's the madam above, on the upper storey, who gives you food. Who is she, anyway? Your mother?'

'No, she passed away long ago.'

'Someone else then.'

Yasmeen nodded her head.

Somewhat sorrowfully, Rafeeq Khan asked, 'Ever get anything else to eat?'

'Of course. I get many other things. Not always, though. Sometimes someone brings curried lamb's knuckles in a bowl from his house. Or if someone feels too loving, he brings fish cooked with rice, of course telling his wife it's for a friend. The next time he says, 'May I have my dish back? The wife's asked for it. I forgot to take it the last time.' '

Both of them laughed.

Rafeeq Khan said, 'Okay, why not do one thing? It's less hot now, and there's really no one around here who knows us. How about covering yourself in a *burqa* and going out with me to have some fun?'

'And what about this rooster?'

'He's yours anyway. Let him stay tied up here. Looks like he's used to you already. Let's first go to the truck stand; there we'll have tea first, and then ... No, I said it all wrong. Let's first go to the place where films are shown. There are restaurants there. We'll first have tea there. The rest I'll leave up to you. You may watch a film or take a walk to the park. Then we can go where the truck stand is and have a feast there. After that I can go home and you can come back here. Today, let it be a break for the lawyer also.'

Yasmeen opened wide the window panels. Outside were the sounds of people walking about and talking on the street below. The breeze coming in was also a little cooler now.

She looked at Rafeeq Khan and said, 'Firstly, evening is the time for work; we're not allowed to go anywhere. Secondly, if I lock the door from outside, the woman upstairs will begin snooping around. She has the duplicate key.'

'No! Really?' Rafeeq Khan said indignantly.

'Oh, yes.'

'The bitch.'

'Whatever else may happen, first of all, she'll untie this one and take him away,' Yasmeen touched the rooster's wings with her toes, as she sat on the cot. 'And then he will be in her tummy.'

Rafeeq Khan shook his head and said, 'In that case, it's useless even to bring *ghee* for you. That too will end up in her stomach. You earn so much, but it seems she's the one who enjoys your share. Look at you! You're practically skin and bones.'

A little later there was a knock on the door. Rafeeq Khan started putting his things together. Yasmeen pushed her head out the opening in the door and whispered something to someone. Then she shut the door and went quickly towards her bed. Lifting her pillow, she picked up the bank note that Rafeeq Khan had given her and asked him, 'Do you have change for this?'

Rafeeq Khan lifted the edge of his shirt and from his waist pocket took out two bank notes of the same denomination as the first one, and put them under the pillow.

Yasmeen stopped for a moment to look into his eyes; then, moving towards the door asked him, 'No change?'

Rafeeq shook his head.

After the person at the door had left, Yasmeen said, 'Now that it's evening, there'll be interruptions like this.'

Rafeeq Khan said, 'I was thinking of leaving anyway. I guess I should visit the lawyer, or my elder brother will get upset.'

'Then take your rooster along.'

'No, no. I told you he was yours. Now your apartment won't look so lonely.'

'And what about his crowing in the morning and waking me up?'

'Well, don't the others—the mullahs from the mosques around—do the same in the morning?' Rafeeq Khan said laughing. Before leaving he took her hand in his and said, 'The next time I come, I'll get my sister-in-law to cook some chicken for you. Not this coming Friday, but the one after. And also some flat-bread fried in butter.'

'Is your sister-in-law a good cook?'

'Oh, yes, yes,' Rafeeq Khan answered. 'Nobody in the world can cook like her. She's herself as nice as her cooking is. If once you see her ...'

Again his sentence remained unfinished.

'If once I see her, what then?'

'Nothing, nothing.'

Going down the stairs, Rafeeq Khan was feeling miserable. Yasmeen peeked out of the door and, trying to cheer him up, quietly said, 'And will this rooster stay tied up to my bedpost even when all my clients come and go?'

'Yes,' Rafeeq Khan said, enjoying the prospect.

'Tell me one thing,' Yasmeen asked.

'What?'

'When you went to that place today, before coming here, where did you keep it? With you, under your arm, or did you leave it at the door? I hear that even the shoes left outside by the worshippers get stolen.'

Rafeeq Khan laughed whole-heartedly. He said, 'Yes, outside, it would have been swiped in a minute. There's a sweetmeat store nearby; I left it inside the store. I know the owner.' Then staying quiet for a moment, as though in thought, Rafeeq Khan said, 'How about going out next time? Will you come out with me?'

As if submitting herself unconditionally to fate, Yasmeen said, 'It's all in hands of the one upstairs.'

—Translated by Faruq Hassan

The Beggar Boy

It was that time of afternoon when beggars came to the door asking for a little flour.

The usual call of the beggar echoed in the alley. The sun was oppressively hot, the air moist and heavy, and the acid of his sweat had begun to sting the beggar boy's nape and underarms. The boy, about twelve or thirteen years old, lifted the heavy sack curtain, peeked inside the door ... and was stunned.

The girl who was kneading dough in front of him didn't move away. It pleased him to look at her, light pink and sitting on the low wooden stool. The boy was dark-skinned; his knees had nearly turned white from being down on the ground so much; and a lock of hair—a *chirki*—longer than the rest of his hair, sprouted awkwardly from the oval of his head like a sudden mushroom rising from an unsuspecting patch of grass.

The beggar boy had been calling at this door regularly for the past few days. At first, they would scold him and chase him away; later, they would throw him a *paisa* or two; but today— today nobody had so much as looked at him even though he had been standing at the door for quite a while.

Once again he repeated his begging call ... to the walls.

'Who is it?' a woman asked from inside the house.

'It's the same beggar boy,' the fair-coloured girl replied.

'Ask him if he is ready to embrace Islam.'

'Hey, you, will you become a Muslim?

The dark-skinned boy gawked at her briefly and moved on without caring to reply. The city was quite far. The prospect of dragging himself all the way to the city to collect his daily crumbs disheartened the boy. If only he had managed to collect some alms at this house, which fell midway between his village and the market in the city, he would have been spared the toils of a long, exhausting walk. But one does not fight over alms!

Coming on to the path, the boy thought, 'What kind of people are they? What has begging got to do with religion?'

Apparently the thought of abandoning his religion was as painful to the boy as the sting of a scorpion.

While the beggar boy was trudging along his path to the city, the children of the family, their play over, came dashing back into the house one by one. They had just come back from watching the wondrous sight of elephants picking up piles of sugarcane and loading them on the empty cars of the freight train. The elephants belonged to the Raja, just as the platform belonged to him. The name of the Raja was also inscribed on two stone slabs on either end of the red-gravel platform. This station was that Raja's *pur*, as in Hastinapur—another city.

But Hastinapur has nothing to do with our story. For one thing, our story isn't all that old; for another, in the days of Hastinapur's eminence, people didn't go about persuading others to change their religion.

The pack of children attacked the food. Back at the station, the wondrous spectacle of the elephants was still in progress, and Ragghu had promised to give them a ride as soon as Bare Babu, the station master, had left for home.

In a small shed not far from the house, the oldest boy of the family, who had arrived from school a week ago to spend his vacation at home, was busy playing a different game with a village girl—a game which we had better not mention in this story lest these sheets might fall into the hands of impressionable children. At any rate, the girl was furious and was threatening the boy. He quickly stuffed a shiny, silver four-*anna* coin in the girl's hand, and the two quietly walked out of the shed, one behind the other.

The girl's skin was taut, smooth and shiny black. She had wrapped her *laihnga* tightly around her legs. Out of the shed, she headed for the railroad tracks, climbed the steep front slope, crossed the tracks, and slowly climbed down the back slope. The boy—his sense of smell still flooded with the girl's strong body odour mixed with the smell of sweat rising from her clothes—was lost in the gentle, tingling music that had risen from her rustic jewelry. Then, after a while, he too started off for home. He was feeling pleasantly hungry.

After everyone had sat down to eat, in walked Bare Babu, the station master. He was a rather rotund man; whenever he laughed, his shirt flapped over his belly as if it harboured an entire colony of leaping frogs.

In a family whose members are all alive and well, whose daughters have not yet been married off and whose boys have returned for vacation—in that family who doesn't have a gargantuan appetite! In Bare Babu's house, just about everyone was a glutton.

As soon as Bare Babu had sat down to eat he called Sanjhli, the third daughter.

In one of the back rooms overlooking the railroad tracks, Sanjhli was standing at the window, her hands holding the bars, her eyes straying beyond the farther slope, her mind searching out a few elegant but meaningless sentences for a piece she was doing for some women's magazine. On the third or fourth of each month, this magazine would arrive by mail in that *pur*. Everyone, even the children, would rush to have a look at it. The manila wrapper which bore the station master's name and title would be carefully unwrapped and stored in the desk drawer. Sometimes, one of these wrappers would find its way— either tucked away inside a book or inadvertently wrapped over something or other—into the house of one of Bare Babu's acquaintances in the village. Mostly, however, it was used as a book-cover.

Sanjhli had somewhat sharp features and a pale-white complexion like *champa* flowers. She had been contributing regularly to this magazine. What exactly she wrote nobody knew. Her pieces, rather harmless and over-sweet and addressed always to an imaginary girlfriend, read rather well. Presently as well, standing by the window, she was trying to fix up some phrases—as usual full of sweet nothings, but mysterious and charming all the same—in her mind. It seemed the stillness of the desolate railroad station had completely absorbed her youthful heart.

She saw the village girl climb up the front slope and disappear behind the tracks, heard her father call her, and left the room to join everybody in the courtyard.

The children hurried through their meal and dashed out, where at the station Ragghu—or rather, the elephants were waiting for them.

The oldest boy was called Manjhla—the middle one. Nobody knew why. The most one could say was that perhaps he had an older brother who had died or disappeared. Anyway, he sat at some distance from his father, fearing lest the colour of his face or the odour of the village girl still clinging to his clothes might give him away. Next to him sat his soft-brained brother who was called Bhauloo the Idiot. The oldest, the middle, and the third daughters sat near the father. The mistress of the house, who had been supervising the meal in the kitchen, was the last to join the lunch. Behind her came Choti—the little one—trailing her loose, baggy pajama pants. Always sloppy, she managed to keep herself dirty and remained unbathed even when she knew a bath was overdue.

Why the girls were called Bari (the oldest), Manjhli (the middle one), Sanjhli (the third one), and Choti (the little one) was something of a mystery, all the more so since the 'little one' had been followed by two other girls. Perhaps because the birth of the last two had not been planned.

Everyone sat down to eat. The subject of the beggar boy with the *chirki* was broached. In the meantime, the beggar boy was dragging his swarthy, emaciated legs along the path to the city market, in the dwindling hope of receiving some grain, a fruit, or a few coins in charity.

The mistress of the house was again expecting. Her skin was the same fresh pink colour as the girls'; the boys, though, had acquired a dark tan as a result of too much running about in the sun.

The mistress of the house asked, 'Did you give the brat something?'

'No.'

'Why?'

'The moment I mentioned about him becoming a Muslim he just bolted.'

Bare Babu felt he had bit on a red, hot chilly. Everyone tensed. In the past three days the family had scarcely discussed anything except the beggar's possible conversion. Will he ... or won't he? The question was so crucial it had preoccupied everyone, big or small, even the Idiot, although it was Choti's considered opinion that the Idiot's feelings ought not to count: he was always too distracted to perform his ritual prayers properly—worse still, he even performed them in the same clothes he had on when he had cuddled the puppy dog.

Although everyone offered the ritual prayers, most did but half-heartedly. The younger boys went to the *idgah* with Bare Babu only twice a year on the occasions of Eid and Baqar Eid; the older ones had to be literally dragged to the mosque for the congregational prayer every Friday. It would seem each one had the uncanny ability to know what was brewing inside the mind of another member of the family. For instance, Manjhla knew very well what kind of man Bare Babu—his father—had been.

Years ago he had once visited Allahabad with Bare Babu. Father and son stayed in the Railway Rest House. In the morning the father went out to take care of some business, and the boy spent the whole morning picking fallen *neem* berries. The father returned past the noon hour and took the boy out to show him around the town. The city was quite big. Throughout, Manjhla could not fail to notice a strange change in his father's attitude— the kind of change which the sons of bad-tempered fathers notice only once or twice in their lives when their fathers unexpectedly become very kind and the sons diffidently try to laugh with them. Anyway, Bare Babu and Manjhla tramped around the city for quite a while.

Bare Babu bought quite a few things: gramophone discs, a thermos, some clothes, magazines for Sanjhli and toys for the other children.

Those toys—they brought back a strange memory. The festival of *Divali* was only a few days away, and the only kinds of toys you could buy were made of either sweetmeats or clay:

men with elephant trunks, six-handed women, weird-looking men mounted on equally bizarre-looking animals with the body of a dog and the head of a tiger, blue-bodied men, monkeys holding maces, and oxen made of coarse brown sugar.

At some point in the shopping Bare Babu suddenly told the boy to stay out and himself took off in another direction. The boy was absorbed in the toys. With one eye glued to a row of toy peacocks, he looked with the other at his father who had meanwhile crossed the street and was now climbing up the stairs of the corner house.

Soon bored with looking at the toys, the boy yawned a few times and then crossed the street and came to the eaves of that house. In one of the shops at the street level of the house, a *paanwari* was deftly applying different watery pastes to *paan* leaves with a pair of wooden sticks.

The boy asked him, 'Who lives upstairs?'

The *paanwari* looked at the boy briefly, then, lowering his eyes back to the paan, said, 'Kali Jan.'

To a child of any other Muslim family the name 'Kali Jan' might have sounded unusual, but not to Manjhla. Long time ago, a Pathan girl called Marjan had worked in Bare Babu's household. The boy right away concluded that this 'Kali Jan' must be a woman. And the realization made his heart beat hard.

With his small legs, which suddenly seemed too heavy for walking, the boy crossed the street at a run. A horse-drawn carriage went past him perilously close, and the whip of the coachman cracked inches above his head. His heart began to pound even more fitfully. He hurried back to the toy shop.

He could scarcely remember when Bare Babu returned and towed him along to the station. But he could remember all too well that his mother tolerated those toys for a day or two and then her patience ran out. She had them thrown out in the alley. Janki's kids gathered the trove and walked away with it. Somebody had asked the soft-brained brother, 'Brand new toys—why throw them out?'

'My mother says they are Hindu gods,' the Idiot replied.

Whenever the girls saw some Hindu worshipper bow reverently before a smooth, shiny stone lying at the foot of a *peepul* tree, they would start to ridicule the poor fellow by mumbling something that sounded like a Sanskrit prayer.

There was not much to do at home but quite a crowd available to do whatever needed to be done. A half dozen daughters, two or three servants, a few relatives who regularly hung around—this left the mother and Manjhli with plenty of time to do their needlework, and Sanjhli to pursue her writing unhindered.

Then suddenly one day the beggar boy appeared on the scene. The first day they scolded and shooed him away—all because he had a *chirki* on his head. Had he been Muslim or at least not had this damnable token of Hindu religion, certainly some charity would have been in order. That a Hindu child had the audacity to come begging at a Muslim house meant that he was woefully ignorant about the religion of its occupants. Or else he was a nitwit who scarcely knew the difference between Hindus and Muslims.

The second day he showed up, it was raining so hard that the alley had become a veritable pool. Dripping wet, he came, lifted the heavy sack curtain, and installed himself in the doorway. Right above the door hung a branch of the banyan tree on which perched a row of crows huddled against the rain. One of the girls asked the boy to come in, but he remained rooted to his place. The girl asked him a second time.

Just then the rumble of a freight train arose at the back of the house. Through the grill of a window, the boy watched the locomotive chug slowly along, its image distorted and trembling in the misty air. Two of the boys began to count the cars in the train.

After the train had passed—and it always took an eternity to pass—the boy begged for a *paisa* or two; he was even prepared to accept a pinch of flour.

A woman came out of the room, swinging her ballooned stomach sideways like a pregnant goat, and asked, 'Who is it?'

'The beggar boy,' replied the girl.

'Why did you let him come in?' the woman asked with some irritation.

The boy started and quickly withdrew to the doorway, where he remained waiting for charity.

A little later a boy came out of the room and handed him a small coin. 'May Ram bless you!' the beggar boy exclaimed and caught on his way out the pregnant woman's angry 'Away, get the hell out of here!'

The next day he was back again, standing in the same spot.

'What is your name?' asked the woman.

'Ramsarna.'

'Ramsarna! Ramsarna!' the children aped, making fun of his name, and laughed.

Choti said, 'Ramsarna!—why, only elephants have such a name.'

'No, they don't,' the other girl observed, rather emphatically.

The male elephants under Ragghu's care had such names as Jhabbu, Masta, Tumbul; the female elephants, names like Champa and Kishori. Even so, the children christened the beggar, 'Ramsarna the elephant.' Choti would often ask, 'Ramsarna the elephant, give us some sugarcane.' And when he would be given flour, the same girl would say, 'Now Ramsarna the elephant is going to bake himself some bread.'

Ramsarna developed a liking for the children. He would come around noon time, stand at the doorway, and ask for a pinch of flour. 'Dhat! Dhat!' the children would make the sound used to urge elephants to kneel. Ramsarna would bring his feeble posterior down, put his begging bowl behind him on the ground, and sit supporting himself on his palms, in the manner of elephants, or so it seemed to him. Often sitting in that posture, he would watch the train go by at the back of the house.

Hearing his begging call for a pinch of flour, the Idiot would throw his hand in the drum, grab literally a pinch of flour between his thumb and the index finger, and thrust it out to the beggar boy: 'Everybody watch! Ramsarna the elephant will eat flour and give milk.'

One day the beggar boy again came when it was pouring down hard. The mistress of the house herself asked him to come in the veranda. The children were jumping around inside the room on cots; the older girls were scattered everywhere in the house; and Bare Babu was away on his job at the railway station. The mistress had a knife in her hand. Her thumb was swathed with a bandage.

'Bahu Ji, you have cut your thumb?' Ramsarna asked.

'Yes.'

'Ram! Ram!' the beggar boy exclaimed in a concerned voice.

A thought flitted across the mistress's mind. 'Why didn't Ramsarna exclaim 'Oh, Allah!'? Or some other expression, such as, 'Oh, Ali!', as some people do?'

After a while she asked, 'What village do you come from?'

'Bhojpur.'

'Is it far?'

'Quite far.'

'You have a mother?'

'No.'

'Father?'

'No.'

'Any relatives?'

'No, Bahu Ji, no one,' Ramsarna said, with touching but feigned ruefulness. He had been wielding this weapon to his advantage for years now and had become quite adept at it. A ray of hope glimmered in his heart. If Bahu Ji could take pity on him, she would make his day.

When the mistress asked him about his kith and kin and his home, Ramsarna realized that beyond a few begging calls he had learnt to parrot, he knew absolutely nothing, or—perhaps—more than that, his life really didn't amount to much.

His mother had died some five or six years ago. Died—as it often happened with low-caste women—of incessant vomiting and diarrhoea after gorging herself on a rotten melon. But who was his father? Where had he come from? And where did he disappear to? Nobody in all of Bhojpur had an answer to these questions. Perhaps he was one of those people who quietly melt away in the crowd of this world without leaving a trace.

Stranger still was the *chirki* which sat on his head like a black bee. Ramsarna perhaps could not even tell how long he had had it and why exactly he wore it.

After the beggar boy had collected the alms—a bit of cornflour—and gone, the pregnant lady of the house sat for a long time thinking about him. She remembered the band of people who had set out for some villages in the east, to spread Islam among people. Missionaries—she knew—who proselytized. Yes, that's it. Why not proselytize Ramsarna?

That night the whole family openly talked about Ramsarna. Bare Babu spoke at length about Aurangzeb's religious policies, with which Manjhla found himself in agreement. The oldest, the middle, the third and the fourth daughters also talked with fervour and excitement. In that desolate, out-of-the-way station—through which most trains sped by so fast they shook the house to its roots—it seemed that finally a mail train had ground to a halt, causing the entire atmosphere to take on the gaiety and exuberance of a country fair.

Slowly the discussion veered to other religious matters. Who used to spread out the prayer-rug on the surface of the river and offer ritual prayers? And who on a bedsheet hanging in midair? As is normal in such discussions, the thought of eventual death crept into everybody. The same wrenching fear gripped them which had gripped the family of Nusuh following the outbreak of the cholera epidemic in that novel by Nazeer Ahmad.

The mother offered her dawn prayer and the younger girls who had gone to bed trembling from fear of the torment meted out in the grave right away sat down to recite from the Holy Koran. Dawn broke.

A couple of times during his work, Bare Babu started. His older brother, the civil surgeon, had developed his own spiritual powers to the point where he could communicate with spirits. For the first time ever, Bare Babu felt the distance which the maddening pursuits of his life had created between them. What was he—Bare Babu? At most a man who unabashedly desired material things. But life's sole purpose was not just to eat, drink and be merry—was it?

The whole of that day Manjhla looked as though he had a lot on his mind. He remembered the time when he was about twelve or thirteen years old and had been caught fooling around with Sukhdaiya, a good eight or ten years his senior. His sister Manjhli had surprised them. She wanted to report the matter to her parents but didn't know quite how to. Finally, she scribbled a few cryptic sentences on a scrap of paper and gave it to her mother, who couldn't make out what the words meant. From that day on, God knows how many Sukhdaiyas had come into his life and, after gratifying his desire for a while, gone away. During this time, the boy did not feel as though he missed them, nor, chances are, did those women themselves feel the need to remember the joys of their brief encounters with their adolescent lover.

Anyway, Manjhla, like his sisters, remained quite distracted that day. And the day after.

About noon, Ramsarna, clad as usual in rags, reappeared. The children saw him all right but remained silent. Only one of the girls came out to give him a small coin and went back in. The mistress also came out of a room, hesitated for a few moments in the veranda, as though thinking deep and hard about something, then she too went back in. The house was gripped by an eerie silence. Even the back of the house seemed strangely still. The older girl breezed in, and then withdrew. The beggar boy hesitated at the doorway for a while and then went away.

He was thinking: 'Strange people! Today they gave me handsome alms but no attention. Why didn't the children feel overjoyed to see me, as they usually did before?'

No sooner had he left than the otherwise dead house became vibrant and alive. The mistress asked, 'Gone—already? I was just coming out to talk to him.'

'No, Mother, you weren't,' the eldest girl snapped. 'You went back in as soon as you saw him.'

'But so did you,' Choti, the ill-mannered girl, retorted.

When Bare Babu returned home for lunch, his very first question was, 'Did Ramsarna show up?'

'Yes,' somebody replied.

'So?'

'So—nothing.' Everyone silently looked around, as if wishing to slink away.

'Well?' he said, a bit shocked.

'He isn't running away—is he? ... maybe tomorrow.'

But the next day, when they suggested embracing Islam, Ramsarna took off like a bullet. True, he was poor; but that didn't mean he would readily give up his faith. So, that's it, then: their charity had a motive! They hoped to convert him!

Many times along the way Ramsarna's feet slowed down to the point where they were barely moving. A spiraling anger seemed to have drained them of all energy. Then he would think with horror of arriving late at the market, and his feet would automatically begin to move faster.

He briefly stopped to wash his face and feet in the river on the way. Then, crossing over to the other bank, he came to a *peepul* tree and prostrated himself in reverential worship before the stone that lay at its foot. For the first time ever, love for his religion had suddenly blossomed in his heart.

When he reached the market, the fair was winding down. People were leaving; those that still remained seemed exasperated by the heat.

That night Ramsarna went to bed without a morsel to eat, just as he often used to before he started begging at Bare Babu's.

The next day he found himself standing before Bare Babu's doorway at his usual time. The children, as soon as they heard his begging call, shouted, 'Ramsarna the elephant is back!' Choti urged him with *'Dhat! Dhat!'* The beggar boy blushed, hesitated a bit and then sat down, supporting himself on his emaciated legs.

'Keep quiet!' the mistress chided the noisy kids. 'Let me talk. ... Hey, Ramsarna, why did you run away yesterday?'

'Oh, nothing, Bahu Ji.'

'Can't be. You left without receiving anything, didn't you?'

Ramsarna remained silent. Two of the children planted themselves on the cot in the yard and began staring at Ramsarna as though he were about to perform some trick.

'Did you get anything to eat yesterday?' the mistress asked.

'Don't you see his face, Mother, he's famished,' the oldest girl observed.

'Who is it?' Manjhli shouted from inside.

'It's Ramsarna the elephant, Baji,' the children replied in unison.

'No, children, no!' the mother chided. 'We don't call human beings elephants and horses! He's a human like you are.'

The children cowered.

Ramsarna was overcome with emotion and began to cry. Nobody in the entire village had ever stood up for him. The other children were forbidden to play with him. Everywhere he went, he was greeted with abuse and driven away. He had never seen people's dwellings from inside. He could not even imagine what their kitchens looked like. And here was this kind woman scolding her own children for his sake! Warm tears flowed down his cheeks to the corners of his mouth, where he licked them off with the tip of his tongue.

'Want to eat something?' the lady asked.

Ramsarna nodded.

One of the girls quickly brought him some bread and the beggar boy hurriedly began to eat it as if he were starving to death.

He was still busy eating when, suddenly, the woman urged him in a solicitous voice, 'Listen, become a Muslim and I will give you nice clothes to wear. Begging is no good. It won't fill all your needs. No home, no place to go, not even a soul to care for you. Stay with us. You can play with the kids. . . .'

'And I will even teach you to read and write,' Manjhli butted in. 'Once you have education, you will become a real babu, a real clerk.'

Ramsarna thought: the girls were really nice to him; Bahu Ji treated him like a mother; all those days of starvation; the chilly nights of Magh; people treating him like an untouchable, or as if he had some incurable disease—not even a whiff of all that in this household! He sat silently listening to the lady's persuasive words.

'So, what do you think? Will you become a Muslim?' the lady repeated. 'Two meals a day—the same food we eat, not some rotten bread thrown at beggars. ...'

This fresh assault of temptation's tidal wave washed away, like some dirt embankment, the last shreds of his hesitation.

'Yes,' Ramsarna consented in a dead voice.

One of the boys made a dash for the station. Within minutes Bare Babu arrived, huffing and puffing.

One of the girls began energetically to work the handle of the water-pump. Ramsarna was now being scrubbed and bathed. Under the cleansing soap his body began to sparkle wherever water touched it. Surprisingly, he wasn't as dark as the dust had made him look.

Inside, the mistress was searching for her younger boys' old clothes. Soon a pile collected on the cot in the veranda.

Ramsarna was tremendously enjoying his bath and didn't at all feel like coming out from under the streaming water. Even his own mother had never bathed and scrubbed him so lovingly.

'Come on, that's enough,' Bare Babu shouted, 'or else he will catch cold.'

One of the girls ran to fetch a towel. Ramsarna was at a loss: he didn't know what to do with this piece of thick, pelted cloth. Just then someone asked, 'Ramsarna the elephant, aren't you going to dry yourself?' And the mistress ordered, 'Hey, dry yourself.'

Ramsarna just stood there, dumbly looking at the towel, when Bare Babu, too, urged him, 'Your body! Come on, dry your body with it.'

Ramsarna began to do as he was told. He looked like the monkey who, acting on the signals of its trainer, occasionally misses a cue and then stands dumbly, not knowing what to do.

The mistress was preoccupied with a thought of her own: the happy occasion called for a *milad*. Why not? And Bare Babu was absorbed in the merry thought that the news of his tremendous feat would even surprise his elder brother.

Although Manjhla had taken no active part in persuading the beggar boy to become a Muslim, his satisfaction was no less: lately he had been somehow avoiding his earlier rakish activities.

Nobody was unhappy in that house that day. Sadness of any kind was out of place in a family whose members were all alive and well, whose daughters had not yet been married off, whose boys had returned home to spend the vacation, and which had been graced with the wealth of sound faith.

Ramsarna was given a pair of pajama-pants and told to go in a secluded corner and change. While he was slipping into his new pants, Manjhla brought him a shirt, and later one of the younger girls, overwhelmed by the spirit of charity, gave him her old pair of black shoes. When he emerged from the corner, the mistress exclaimed, 'Now you look like a prince!'

The beggar boy found himself standing in the veranda— dumb, lost, uncomprehending. He could not think of anything to say. He was overwhelmed, like everyone else, by so much excitement. But the reason behind the excitement—everyone seemed to have already forgotten it.

The mistress grabbed the boy's hand and brought him to a low, wooden stool. As he was being seated, somebody pulled him up again; he was seated only after Choti had brought a prayer-rug and spread it on the stool first.

'Now, there—make a Muslim out of him!' the mistress said to Bare Babu.

'Hunh?' Bare Babu took a few steps forward, then backed off, not knowing exactly what to do.

'What are you waiting for?' the mistress demanded.

'You are doing the right thing. I'll do my part tomorrow.' Bare Babu looked at his wife as though he had thrown a riddle at her to solve.

The mistress asked the boy, 'You are becoming a Muslim out of your own free will—aren't you?'

Ramsarna remained silent.

'Come on, say that you are. Aren't you?' The mistress shook his shoulder.

'What, Bahu Ji?' Ramsarna asked.

'I'm asking you to say that you aren't accepting Islam under duress—right? Look, nobody is threatening you, nobody is strangling you either ... you are converting happily, right?'

'Out of your own free will,' Bare Babu chimed in.

Ramsarna indicated with a nod that it was so. But everyone asked again, 'Say it!'

'Enough!' Bare Babu said, 'you have asked him three times already. That'll do.'

'From today your name is Khuda Bakhsh.'

Ramsarna shook his head in affirmation. To having a new name he had no objection at all.

'Now we must have Bakhshu recite the *kalima*,' one of the girls said, 'Come on, Bakhshu, say *La ilaha. ...*'

'*La ilaha. ...*'

'*... il-lal-lah.*'

'*... il-lal-lah.*'

'No, that won't do,' Bare Babu butted in. 'First put a cap on his head.'

'And the *chirki*?' the Idiot had a brainstorm. 'How can he become a Muslim with his *chirki* intact?'

Yes, the *chirki*? How can he? The words were charged with magic. The soft-brained brother had hit upon the most important thing!

One of the girls dashed off, returning seconds later with a pair of scissors; and the oldest boy moved forward to clip off the *chirki*. Suddenly the beggar boy threw both his hands over his head and closed them tightly over that small tuft of extra-long hair, as though he was guarding with his life his sole asset. If he lost it, he would truly become a pauper.

'Hey, let go of it,' Manjhla chided. 'Let me get rid of it for you.'

'No, no. I won't let you do it,' the boy broke into tears.

'Er-r-r ... what's the matter with him?' the mistress said in disbelief.

A real tug-of-war broke out over the beggar boy's head. Meanwhile the boy had started to sob violently.

'How are you going to become a Muslim if you won't get rid of the *chirki*? Damn it, you will remain a beggar for the rest of your life.'

'I won't let you cut it,' said the boy firmly.

'I'll see how you won't. I will. ...' Bare Babu fumed.

'*I won't let you.*'

Hell broke loose over the boy's head. A lot of tugging and pulling went on for quite a while, then someone pushed him off the stool, and another shouted, 'Get the hell out of here.'

The beggar boy picked himself up and moved away with feet drained of all energy. He passed by the hand-pump where his torn shirt lay crumpled. At the door he picked up his begging-bowl. Just then someone shouted from inside, 'Grab him!'

Ramsarna leaped out of the door into the alley.

Back in the house everyone was thrown into a wistful mood as though they had just returned from a funeral. Bare Babu went back to the station; the younger kids went out to put stones on the train tracks; the Idiot took out his collection of empty cigarette packets and began to count them; and Manjhli noticed that Manjhla was again climbing up the slope way down by the tracks.

Manjhli was suddenly overwhelmed by an urgent desire to write a letter addressed to her imaginary girlfriend, describing the loneliness, the silence that pervaded her surroundings, broken only by the occasional rumble of a bullock-cart as it passed along the path and then silence would settle in again like a surface of water disturbed and then calmed.

Somewhere along a narrow path the beggar boy was immersed in a thought of his own: What had happened with him? If it was all a joke, why did he panic and take off? But if it was not a joke and meant really abandoning his religion, then it had been a narrow escape indeed, and he must thank his stars for that.

He didn't feel like returning to his village. His new clothes seemed to please him, and after the refreshing bath he was also feeling nicely hungry. He fished around for something to eat in the vegetation on either side of the narrow path. Finding nothing, he set out in the direction of the city.

The sun had all but set by the time Ramsarna made it to the city. He had scarcely called at two or three doors when it got quite dark. He settled on a jutting raised porch outside a shop and began to munch from the small leaf-basket which a sweetmeat seller had given him. The contents were mostly sweetmeat-crumbs, stuck with ant-legs, and had somehow become mixed with a bit of salted yogurt.

The sky, turning a flaming red for a few moments, became completely dark. The worshippers who had come for the evening service in the temple across from where he was sitting returned home. The priest came out and sat down on the elevated, flat unroofed porch outside.

The air was stuffy. The priest looked like a dark statue. A bird flapped its wings and flew above Ramsarna's head. Feeling bored and lonely, the boy got up and walked over to the priest. Since afternoon he had been feeling rather restless, needing to unburden himself about what had happened.

'Who is it?' the priest asked in a stern voice.

'Maharaj, it is me, Ramsarna.'

'So?'

'Nothing. I just came to pay my respects.'

'Where do you live?'

'In Bhojpur.'

'A beggar?'

'Yes.'

'Have parents?'

'No.'

'Sit down.'

Ramsarna sat down.

'Your caste?'

Ramsarna remained silent. So did the priest.

After a while, twirling the *chirki* around his finger, Ramsarna said, 'Maharaj, there is a house on my way. Muslims live in it. A man there wanted to cut off my *chirki* today.'

'Who's that scoundrel?' the priest asked.

'He's called Bare Babu.'

'So did you let him do it?'

'No, Maharaj.'

'What else did they want?'

'They said, "Become a Muslim. Stay with us. We will give you clothes, and you will have plenty of food to eat."'

Ramsarna took a long time relating his tale of misfortune, which the priest every now and then punctuated with cries of 'Scoundrels! Rascals!' When Ramsarna had finished, the priest gestured for him to come close and began examining his *chirki*. Then, holding the boy's hand, the priest praised him for a long time. In his opinion, Ramsarna had performed a meritorious deed which only members of the highest caste were capable of. 'Tomorrow, I'll relate this story to everyone,' he said. 'I'll tell them how such-and-such a child valiantly defended his religion. I'm sure some noble soul or the other will take you in.'

His belly full, Ramsarna fell fast asleep in the temple courtyard. Suddenly he felt someone's warm breath over his face. Fear gripped him. In his confusion, he tried to jump up, but felt as though his slight, emaciated body had been pinned down by a heavy stone. He tried to scream, but the priest pressed his hand down hard on Ramsarna's mouth. The two tangled. Ramsarna had no doubts about the priest's intentions. Like every orphaned, indigent boy who tramped the streets begging, he had come to know, despite his rather young age, life's many unpleasant secrets.

Finally, the boy succeeded in wrenching himself free of the priest, just as a snake shakes its head free of the grip of a mongoose, and dashed out of the temple. He kept running madly for quite some distance, though he didn't have to, for the priest thought it pointless to pursue him.

Roaming around for a while Ramsarna found himself in front of a cattle lock-up. This was the last brick building at the city's

outermost edge. A river flowed by it and the bridge spanning it
led toward the path to the village. Ramsarna, however, didn't
find enough courage in himself to return to his village so late at
night. Instead, he climbed the wall of the cattle lock-up and
jumped down into the compound where he spotted a donkey. In
the darkness the donkey looked as though it were sleeping. A
little distance from the animal, Ramsarna laid his tired body
down on a heap of hay and dozed off.

The next morning as he was walking back to the village,
Ramsarna thought: why did those people want to make him a
Muslim? What would have they gained by that? They didn't
love him; and they certainly wouldn't have fed him forever.

Then the thought occurred to him: the priest was no
well-wisher of his, either. Why, then, did he want him to remain
a Hindu? What good would it have done him? Those other
high-caste Hindus—they wouldn't have taken him in.

After thinking long and hard, Ramsarna concluded: let's just
say there are people who do many things without a reason. Bare
Babu and his family were like that. So was the priest. And so
had been his own father earlier.

—*Translated by Muhammad Umar Memon*

The Cactus

After ringing the doorbell, my wife and I stood quietly beside the door, each thinking the same thoughts: let's see how this home owner turns out, what his wife is like, and, above all, what this house looks like from inside.

On the outside the house was very quiet—almost soothingly tranquil, something that we had been looking for. A cool morning breeze was blowing through the foliage. Beside the rustle of leaves, only one other sound could be heard—that of music coming from somewhere far off.

Near where I stood, there was a full grown cactus. Somehow, it hinted at the personality of the owner of the house. Some time ago, when this cactus had grown past the window and its top had begun to touch the eave, a hole had been cut into the overhanging eave to allow it to grow unimpeded. Now its top had gone through the hole and stood above it proudly, looking somewhat like an ostrich, holding his head high after he has beaten back his rivals.

Gesturing with my eyes, I pointed out the hole in the eave to my wife.

'I've noticed already. Looks like they're nice people.'

We heard the distant rattle of a door latch being opened somewhere inside the house. Then the voice of a middle-aged man inquired: 'Who is it?'

After a short pause I said that we had come to have a look at the house. The voice asked, 'Are you from some real estate agency?'

'No,' I answered.

'Okay. Just a minute.' These words were followed by a long silence. The tension that usually precedes such a visit had disappeared. In a sense we had been introduced to the owner of the house. So we felt free enough to walk about a bit and look around the house. The information we had been given about the house was right: it had been built on an area of about 600 square yards and had two stories. We could also see that it had been well-maintained. All the window panes were intact. There were no drips of paint on the glass from when the wood had been painted.

Our probing had gone only so far when the front door opened and the owner of the house came out and stood on the doorstep. He examined us cursorily and asked us to come in.

First my wife entered the vestibule. I followed her, with the owner behind me. Although it was 10 a.m., the house still seemed asleep. In the drawing room, the owner asked us questions about where we lived, what we did for a living, and so on; then, telling us to wait for a few moments, he vanished into the corridor that separated the rows of rooms. The two of us tried, once again, through eye gestures to tell each other our impressions.

The room was remarkably tidy. The curtains were separated in the middle and neatly tied with ribbons. Pictures hung on the walls everywhere and, astonishingly, there were no cobwebs behind any frame, nor even any dust on the glass. The cream coloured back-covers for the sofa looked fresh and clean, as if they had been placed there just that morning.

On the other side of this large room, behind the glass bead-curtain, was the dining table. Extra chairs had been lined up nicely along the wall.

'They don't seem very anxious to sell,' my wife whispered.

'Why?' I inquired, startled.

'All the signs show that they're well-nestled. Those who want to sell often just cosmetically dress things up. They don't polish the leaves of the rubber plant in their drawing room, do they?'

I had missed that one.

That gentleman came into the dining room through some secret entrance and stood parting the glass bead-curtain. 'Let me show you around the house,' he said.

'Wouldn't it be better if we had some idea of the price first, to see if we can afford the house?' I said.

'Don't worry about that,' he said. 'Have a look at it first.'

'But we might just end up tiring you for nothing if we ...' said my wife.

'...If you can't buy the house?' he completed her sentence. 'Don't let that bother you. This has been our routine for quite a few days now. Interested parties come to look at the house; some object to this, some to that. But neither my wife nor I feel tired of that,' he added.

'No,' my wife said, 'what we mean is that it wouldn't be fair if we made you take the trouble of showing us the house but later couldn't even come up with enough for the down payment.'

'There's another possibility too. This house may not prove to be worth the money you want to spend on it,' he said with a laugh. 'Come on, madam, let's start with the ground level.'

'All right, as you wish,' my wife answered.

'As you can see, there are three bedrooms at this level—one master bedroom and two others which are almost as big as the master bedroom. Each has an attached bathroom.'

My wife whispered in my ear, 'Perhaps the other buyers found the price too steep.'

'Yes, maybe. But there's no harm in looking at the house,' I answered in a similar whisper.

From the construction and size of the house we had already gathered that it was going to be way beyond our means. The owner walked ahead of us, urging us to go inside every room and look at it. Every room was nicely decorated. Photographs of the members of the family, old and young, hung on the walls. The old members in black and white, the young almost all in colour. There were also a few pictures of some foreigners. The kitchen as well as the gallery had pictures on the walls, and there were flower pots everywhere. My wife observed the American-style kitchen with great interest and attention. I could easily guess what was going through her mind: in our own house, whenever we build it—if we don't buy this one—the layout of this kitchen will come in very handy.

Then we started going up the stairs. Like those downstairs, the bedrooms upstairs were also unoccupied. There was no one

asleep in any of them. In one room there was a child's bike and
a rocking horse on the carpet.

'This is perhaps your children's room?' my wife inquired.

'No, my children's children,' the man answered good-
humouredly.

'Are they at school?' I asked. But as soon as I had put the
question, I realized that it was foolish because the day was a
holiday.

'No, the two children whose room this is have gone to Kuwait.'

After admiring the neatness and order of the room for a short
while, my wife said, 'And it looks as if the things were put in
order just today after the children left for school.'

'Or,' I said, 'the children are so well-mannered that they
themselves put everything in place before they left the room.'

'Oh, no,' the man answered, 'I could tell you volumes about
their good manners! Even when they know they are leaving for
Kuwait, they leave a mess behind on the rug, as if they're going
to be back in a short while to resume their play. It's we who
have to arrange their things after coming back from the airport.'

It took us quite some time to go through that room. When we
came out we ran into the owner's wife. She was watering the
plants in the flower pots hanging in the corridor. On seeing us,
she said, 'Forgive me. I was on the top floor when you people
came in.'

My wife whispered, 'Oh God. We've yet to see the top floor!'

The owner's wife put the watering can down on the floor and
wiped her wet hands on one end of her head-scarf. We stood
there in awkward silence. Before long the gentleman spoke:
'Let's go up to the top floor.'

We both followed him up the stairs, me going willingly, but
my wife quite against her wishes. Her knees bother her when
she climbs stairs. It was, thus, no surprise to me that after
climbing a couple of steps she said, 'You go ahead. I'll stay
here and wait.'

She went towards the owner's wife, took the watering can
from her hand and said, 'Allow me to do this. How many
children do you have?'

The top floor—actually the roof—had a big open courtyard, on one end of which were two large rooms. In front of them was a large, L-shaped wooden patio with a banister. The flower pots on the patio had been watered recently. It seemed that the owner and his wife did not come to this level very often, perhaps only when the plants had to be watered. It took the owner quite some time to open the locks; they were somewhat rusted.

As I entered the first of these rooms, I asked the owner, 'And whose room is this?'

'My elder son's,' he answered.

'Where is he?'

'In the States. In Houston.'

Mattresses, without any coverings, lay on the box-springs. Various things had been piled up in one corner and covered with a tarp. One shelf was filled with books; the other with souvenirs from various countries.

Leaning against the railing on the patio, we talked for a long time about the world's affairs. He asked me if I needed the house as my principle residence.

'Yes, at least that's our wish. Only God knows when that wish will be granted.'

'For how many people? I mean how big is your family?'

'There are three children and two adults.'

'Then this house is probably going to be a little too big for you.'

'Yeah. I too think so. How many children do you have?'

'Three daughters, two sons, and seven grandchildren.'

'I don't see any of them here,' I said, a little impudently.

'Oh, one daughter is in Kuwait, the other two in Canada. One son is in the States and the second one in England.'

Suddenly he seemed an old man to me. I noticed that near his earlobes some hair had escaped the blade. In some other spots on his cheeks he also hadn't shaved carefully. There were white rings around his pupils in both eyes.

When we came downstairs, his wife and mine were sitting side by side in the drawing room. In front of them, there was coffee and some dry fruit on the trolley.

I asked for his permission to leave.

He said, 'So, did you like the house?'

'Very much,' I answered.

'But ...' my wife intervened.

'I know. You cannot take it. It's too big for your needs,' he said. 'But that doesn't make much difference, does it? We can still have a cup of coffee together.' He sat down on the sofa and started to break almonds with a nutcracker. Willy-nilly, I too sat down. His wife was serving coffee to my wife.

When, after coffee, my wife and I came out of the house, his wife accompanied us to the door. She said, 'Come, let me show you my garden.'

'We've already met your cactus.'

'Oh, there are many more; meet them too.'

In a small garden she had accumulated a number of different shrubs and trees and creepers. There were a couple of palms as well. They all seemed to have been watered that morning.

Before leaving we stood in their car port, talking.

'Where would you live in Karachi after selling the house?' I asked him.

'Don't know yet,' he said.

'With some relatives? Do you have any here?'

'Yes, there are some. Here and there. But it is useless to count on anyone's being near you in order to live.'

Unexpectedly, his wife joined in the conversation. 'We had chosen this city—that is, Hyderabad—and, more particularly, this house after a lot of consideration. Now we're afraid of going to a new place. Who knows who might be where tomorrow?'

On the way back, my wife, as soon as she got in the car, said, 'Do you know they have three daughters?'

'Yes, I know.'

'And two sons?'

'I know that too.'

'They're all married and gone, leaving the nest empty for the old folks. During the holidays, sometimes a daughter or the other visits them, but that's all.'

'What about the sons?'

'One is married to an American. She works in the States. She sometimes comes to visit them. The other son's wife is a Pakistani. She doesn't get along with the old lady.'

I looked at my wife admiringly. I couldn't have got that sort of information out of the old fellow.

—Translated by Faruq Hassan

A Requiem for the Earth

Eventually there came a time when persons, keenly interested in statistics, noticed that the matrimonial ads appearing in the newspapers mostly asked for suitable wives. The demand for husbands diminished day by day and in the end ceased altogether. The text of the ads also underwent a change. People became less demanding. At first they were fussy about the girl. They wanted her to belong to a certain place or caste or religion or to a particular religious sect. But gradually they grew less inhibited and expressed their readiness to marry a woman from anywhere. Other conditions, however, were still in force. For instance, the girl had to be a Roman Catholic, a Shia, or Brahmin, and even when she moved abroad following her marriage it was only to join one of her fellow countrymen.

Then the young men seemed willing to marry anyone. It did not matter if the girl happened to be a Christian or a Muslim or a Hindu. Gone were the days when the girl had to be strictly a Maronite Christian or a Bohri Muslim or a Chatterji Brahmin.

It seemed as if the boys and their parents were becoming less and less fastidious as far as the would-be wife or daughter-in-law was concerned. Nevertheless their greed to acquire the latest electronic devices or accumulate creature comforts increased sensationally. Another encouraging change was noticeable in the wording of these ads. The girl no longer necessarily had to be fair or handsome. Ultimately it was enough if she happened to be a girl. Formerly it was customary to mention the required age of the girl in the very first line of the ad. By and by it was relegated to the last line, only to be completely omitted later on. Now nobody looked for a girl who said her prayers five times a day, fasted regularly during Ramzan, and was handsome, willowy, virgin, highly-educated and of wheaten complexion. Nor was it necessary for her anymore to be between twenty and twenty-five years of age. All these conditions were becoming obsolete.

But the statistics hounds were stunned when they came across ads by wife-hungry Dutch and German boys in a Malayalam newspaper. Other boys, from as far away as Trinidad and America, began to take out ads captioned WANTED A WIFE in Urdu papers. A number of marriage bureaus were set up which worked on a global basis. People imagined that the bureaus had something to do with the WHO or the publicity wing of the UN. It appeared very likely that these institutions were trying to mould the various nations of the world into a universal brotherhood.

By now there appeared in English and French newspapers group ads of North American Indian and black males looking for wives, and even in Urdu and Hindi newspapers one could see lists of men belonging to practically every country on earth who wished to get married. It looked as if the world was facing a serious shortage of women.

Eventually even those demands which adhered to the dictates of race and religion were brushed aside. Such demands were the most obdurate of all but ultimately men appeared willing to give them up as well. For instance, no one objected if the girl happened to be an untouchable, and in some papers one even came upon ads like this one:

WHITE CITIZEN, BELONGING TO SOUTH AFRICA, AGE ABOUT 45, CONNECTED BY BIRTH TO THE DUTCH REFORMED CHURCH, WELL-OFF, WANTS TO MARRY. THE GIRL NEED NOT BELONG TO ANY PARTICULAR RACE OR RELIGION.

These events took place at a time when women had become familiar with the pictures that appeared in the newspapers, showing six or even eight newborn babies lying in a row, wearing clean white diapers or clouts. All of them belonged to the same mother. Some of the papers carried a photograph of the mother as well, who appeared to be laughing, or rather making a half-hearted attempt to laugh. The photographer must have insisted: 'Laugh a little, please.' Some of the women remarked that the mother invariably looked very much down in the mouth. And those among them who were a little outspoken said impulsively: 'Is she a woman or a bitch? A litter of ten, neither more nor less.' After looking at one of these pictures a son asked his mother: 'How many babies did this woman have?' and when the mother said, 'Nine or ten,' he exclaimed in astonishment: 'Wow, even more than our doggy.'

Now the scientists took over from the columnists in the newspaper. All that the columnists were required to do was to touch up what the scientists said, turning it into something the general reader could follow. Genetics was the rage everywhere. Even the villagers were no longer unfamiliar with expressions like 'genes' and 'chromosomes' and often asked about 'X-Y' and 'X-X' to find out what these meant. Everyone wanted to know why girls were becoming scarce in the world, or why babies, born either in the natural way or after taking fertility pills, always turned out to be male, or why when a litter was brought forth by a mother the babies lacked viability. Even 'viable,' a new buzzword, became old-fashioned or overfamiliar. Everyone knew that a collective grave would be needed for the litter and there would be no burial rites. The reason being that after their birth these babies lay completely still and made not the faintest of noises. Pink like the newborn of cats or rabbits, hairless, eyes closed, there was something slimy about them. They did not resemble human babies at all.

The genetics experts had cried themselves hoarse saying that for some reason, which could not be determined for the time being, women had lost the capacity to give birth to female babies. And this was later on confirmed following the study of the chromosomes of women who had volunteered for scientific experiments.

Such were the times when, during a get-together, at which specialists in child therapy were also present, the participants began to discuss why mothers were constantly complaining that their children had become averse to food. The mother had to wheedle and coax the child and only then did he condescend to take a bite or two with a wry expression. The mother wanders after him, a plate in her hand, and when the child keeps on saying 'No, no,' she sits down ruefully, wondering how on earth her child would grow up. There was a view, which was thought worthy of serious consideration, that since mothers often stay away from their homes, they force-feed the children in order to play down their feelings of guilt and also to atone for their sinfulness.

One of those present said: 'Perhaps the women are trying to make up for the frustrations they knew in their childhood.'

His view also was regarded as weighty. But then someone said in all innocence: 'How strange it is. When I was a child every child had only one thing on his mind and that was to scrounge around somehow in the cupboard, the storeroom, or the *chhinka*, looking for something to eat. He was roundly rebuked for trying like mad to stuff himself with food and now ...'

At this point someone chipped in to say: 'So in your opinion all this transformation has been caused by changed circumstances?'

The man tried to say: 'What I mean is that in the past ...'

The one who had interrupted him said bitterly: 'What you mean is that only the days which have gone by were good simply because the kids took their meals without being told to and now they don't do so even when food is pressed upon them.'

Although the man spoke seriously his tone betrayed a hint of embarrassment. 'Excuse me, I have nothing to do with the sciences. That is why I have to fall back on commonplace words. Maybe the children of our times have stopped eating because nuclear tests are being conducted and rockets sent into space every other day, nuclear devices keep getting lost in the mountains and the oceans, there is radioactive fallout, nuclear waste, the spraying of insecticide on crops, food preservatives, monosodium glutamate, etc., and the presence of all these things in the atmosphere, in the water ...'

All those attending the party suddenly fell silent and quietly began to sip their cognac and whiskey.

The man who had raised the objection still sat facing the simple-minded person and had no idea how to shake him off. In the end he raised his cup to his nose and began to sniff at it.

The simple-minded person emptied his glass, placed it on a tray carried by a waiter who happened to be around, bowed to the company and said, 'You will excuse me,' and silently walked away to join another group.

* * *

A further change ensued. Once you could see, both along the seaside and in the marketplace, girls attired in brilliantly coloured saris, skirts, sarongs, and *kebaya*, walking about with their boyfriends; and young women came out in the evening to stroll around pushing their baby carriages. Now you only saw middle-aged and old women. They wore floppy dresses in muted colours, as befitted their age. The dresses were baggy because the women were already past the age at which the body is worth showing off.

These women became tired after walking only a little distance and threw themselves on the benches. Young men went past without the least bit of curiousity and spent most of their vacations fishing. By now fishing had acquired the status of an art because fish had become rare in the sea and were equally hard to come by in the rivers and lakes. The poisonous effluents discharged by the factories and the oil spilled by giant oil tankers which had sunk sufficed to bring marine life to a state of near extinction. In any case, birds, animals, and creatures of the sea, not being as tough as men, were not able to cope with these new disasters.

At first, love poetry came into prominence for quite a while in every language. Later on the poets tried to tackle more complicated themes and in the end put their pens aside. The number of stories appearing in magazines decreased steadily and very few films were being made. As far as adventure stories were concerned, writers belonging to the older generation regularly showed a girl accompanying the man setting out on a quest. But the girls had already vanished. So how could a writer put one in his story?

Even for the stories involving Tarzan you needed a girl some gorilla or other could kidnap. But as there were not enough girls to go around for men, how could you find one for the gorillas? In the stories men usually had to struggle very hard to win someone's hand. Sometimes a woman became the sweetheart of an outlaw and was killed alongside him in an encounter with the police. All these matters had become quite dated and the new generation, in any case not very new anymore, watched these old movies with apathy.

At the cinemas, whenever a scene appeared on the screen showing a man and a woman in an intensely emotional mood, the young, belonging to the new generation, expressed their boredom by yawning vigorously. Ultimately most of the cinemas closed down and the governments turned them into ammunition depots. Film studios were used as laagers for tanks. The men representing the older generation were dismayed by the younger people's attitude. 'Let us assume that women will start giving birth to female babies tomorrow. Would these fellows be able to find the girls attractive or not?'

Indeed, there was a strong possibility that girls would be born again. Every major research center was investigating the causes of this global disease. Every now and then countries like Japan, China, the US, Russia, Great Britain, France, Switzerland, and Germany announced that the results of a new drug, which was being tested on animals, were encouraging. For instance, there was a cow in Argentina which was producing only female calves in great profusion, and that too every three months. Two different propitious news stories came out of some countries at the same time.

It was reported that an illustrious Hindu saint or a Muslim holy man, lost in divine ecstasy, gave a handful of dust to a woman. She swallowed it and gave birth to a girl. However, the girl would have to be guarded and kept in seclusion so as not to expose her to any kind of eclipse, solar or lunar, and also to deny men the opportunity of looking at her, since men's lustful glances can be as baleful as an eclipse. The other heartening news was the development of a new drug at a certain research center. With its use the ratio of males to females in the second generation of rats changed to 1:2, in the third generation to 1:3, and by the time the tenth generation grew up the she-rats were giving birth to she-rats only. Once again the columns in the newspapers were crammed with what the scientists had to say. True, the real drug had still to be discovered but what a great achievement it was that if you gave a new product called j7314 to female dogs during the first week of the estrus cycle you could make them give birth only to males, and if the product was given to them during the last week of the cycle, corresponding to menstruation, the litter consisted entirely of females. But there was a snag. It was reported that the administration of the drug led to the appearance of cancer in the dogs' teats and therefore it wasn't proper to test it on women.

The older people appeared somewhat bored, as if fed up with all such news items, and felt that the world was approaching its end. A great many of them shouted in their sleep and said by way of explanation: 'The mountains are flying' or 'A great hole has appeared right through the globe and as a result the waters of the Ganges-Jumna are pouring into the Missouri-Mississippi, and those of the Missouri-Mississippi into the Ganges-Jumna.'

The men belonging to the next generation, who still nursed desires of marriage, read every news report with great interest and quarreled among themselves without rhyme or reason. During the day most of them smoked cigarettes loaded with hashish, and drank an infusion of boiled opium capsules and tea leaves at night. Thus one could now see everywhere lovely fields sown with opium, and nobody found its cultivation objectionable.

But the third generation had nothing in common with the first two. Those belonging to it had rarely seen a woman and when they did come across one they only saw somebody who was either toothless or had fitted herself out with dentures and whose eyes appeared lifeless behind round glasses. Who on earth would care to write poetry after coming face to face with such a creature?

At times someone would callously exclaim: 'This old hag! Did the heartbeat of millions go up on seeing her? I can't believe it.'

On hearing such an expression of disbelief an old man once remarked: 'Look here, when we were your age and our parents enlarged upon the scenic marvels of the countries they had emigrated from and spoke of rivers, of hills covered with grass and forests and the animals which lived in them, we also used to laugh. Their recollections seemed so far-fetched to us.'

The young people found classical literature merely exasperating. It was left on the shelves in the libraries to be consumed by the termites. Old films which lay unused in the cans disintegrated and turned into lumps. The issue of women's survival or non-survival meant nothing to the young men because they had grown up playing among boys. They had no idea whatsoever what a sister was like either as a playmate or a companion.

The pace of the research work largely reflected, in a subservient way, the speed at which the women were disappearing from the face of the earth. So long as the number of women, particularly of those women who could still be expected to give birth to a child, was on the decrease, the governments had no choice but to divert most of their fiscal resources away from the establishments for space research and planetary exploration to the 'Save Mankind' fund.

There was no end to the articles on radiation, chromosomes, and genetics in the papers. The scientists who contributed these were also busy vilifying each other. They were of the opinion that if a certain experiment hadn't been carried out in a certain country twenty years ago, the present predicament could have been averted. The country in which the experiment had been conducted twenty years ago countered by putting the blame on the first country and said that the initial step in this direction had been taken by it fifty years ago. The columnists, whose syndicated writings appeared in every country, behaved with moderation, expressing the view that it would be more appropriate to say that we still don't quite know why women have fallen victim to this particular type of sterility. Therefore as long as we were unable to pinpoint correctly the reason or reasons for it, every judgment tended to be premature and the cure a little further off.

A man of letters, who had given up writing nearly fifty years ago, while leafing through his old diaries, came across a strange dream he had once had. He had written about it at length:

Very hot today. While reading the newspaper after my lunch I fell asleep. I saw an island in the distance on which there was a clump of coconut palms. In the background the sky was absolutely clear, looking as though a washerman had put an excess of blueing on a white sheet. I was on a fishing boat, far away from the island. A light wind was blowing. Somehow I had intuited that we were in the Pacific. My companions, who may have been either Japanese or Filipino, were busy hauling in the fish at the other end. I was looking at the island. And then, as though out of thin air, someone said to me: 'That's the Bikini Atoll.' And immediately afterwards there rose a column of smoke from the coconut palm grove. The island vanished behind a pall of smoke. Ash began to fall on us as it must have fallen long ago on the city of Pompeii. My companions jumped into the sea although it was infested with sharks. I tried to crawl under a plank as the boat, buffeted by the waves, began to founder. When I woke up, I sensed that some of the ash had settled on one of my thighs also. It seemed to me as if I were experiencing a burning sensation where the ash had fallen.

The writer found it annoying that he should have missed out on the news value of so significant a dream at the time.

He mailed the dream to a magazine for publication. In a letter which he appended to his dream the writer wondered whether the present situation was the outcome of such and such experiments conducted over the years on certain islands and in certain deserts. The tests had been carried out so secretly that even those who lived nearby didn't know anything about them.

Every week, for six months, he waited for his letter to appear and in the end came to the conclusion that perhaps the opinion he had formed fifty years ago concerning the newsworthiness of his dream was a sensible one.

Finally it came to pass that people woke up to the realization that women had already disappeared from the world. The inhabitants of every village, every hamlet, could tell you when and where its last female member had passed away and where she lay buried now. Even in cities, which boasted of populations of ten million, people could recall the names of the last five or ten women who had died there. Sheila, for instance, and before her, Jodha, and before her, Sakeena, and before her ...

And these names could be rattled off by just about anyone at all. However, ascertaining these names was almost as useful, or as pointless, as learning the name of a bird which had become extinct. Take the dodo, for instance, which could still be seen in Mauritius as late as the seventeenth century. Who on earth would now care to find out where the bones of the last dodo alive lay buried?

The announcement by the Save Mankind Center that a cure had been found led to an intensification of the search for the surviving women. When people came to know that after taking the medicine every woman would be able to have children the normal way, and some of them would be male and some female, the terror of extinction served as a spur. The quest took on a sense of urgency because no one wants to die in obscurity. The fear that the cities would stand as they were but be empty of the people who built them was a nagging one. Everywhere men could be seen on the move, either individually or in groups, trying to track down women. All day long aircraft flew over rural areas in the hope that a woman might be found among the cotton-pickers or even among those who collect firewood in the jungles and carry it home. The ads which appeared on the radio and the TV or on posters were identical: where are the last women? a cure has been found. But as the days went by dejected

people began to make their way back home from the islands
and the oases. And the staff of the world organization called
'Save Mankind' became increasingly panicky.

Then one day they received a letter which had been mailed to
the organization from some remote mountain village. It said:

> In a nearby village I have seen on a number of occasions a woman
> walking towards a hill, accompanied by her husband. He is a former
> schoolmaster and now herds sheep but spends most of his time
> looking after his wife. Both their young sons died suddenly of a
> blood disease. I know where the woman lives.

Early the next morning several helicopters landed
simultaneously on the hill mentioned in the letter. It had a lake
on the top and was flanked by a glacier on one side and by a
large plain on the other. The sheep bolted as the helicopters
came down noisily and someone peered out from a cave at the
newcomers.

Some men got down from the helicopters and headed for the
cave, making their way across the plain. Their boots kept on
trampling small lilac flowers hidden in the grass which had
barely raised their heads in the morning breeze to sun themselves
in the light of the day just beginning.

Having reached the cave they called out to the man in several
languages and tried to attract his attention by clapping their
hands. But complete silence prevailed inside the cave.

When one of the visitors tried to look in by raising the curtain
made of animal skin, the man hiding inside said testily: 'What
do you want?'

'We have come here on behalf of the global organization
called "Save Mankind".'

The man inside asked in a voice which sounded even more
angry: 'What for?'

'Did you hear the announcement—about the last woman? Do
you own a radio?'

After a while the man replied: 'Yes, I have a radio, and I
have been listening to the announcements for a long time.'

'We have come here with a medicine,' the interpreter said. 'For your wife.'

'We don't need any medicine,' the man said indignantly.

The men outside tried to soften him with sweet-talk. If only, they said, he would let them try on his wife, just once, for the sake of mankind's survival, the medicine they had brought with them. She represented the last hope for everyone.

He raised the curtain, came out and sat down on a stone facing them. The woman also emerged from hiding and seated herself on a stone near the cave's entrance. She had very little life left in her. Both of them appeared fearless.

At last, looking at his wife, the man addressed his visitors in a disagreeable tone: 'Absolutely not. For me the most important thing is my wife.'

'Not the survival of mankind?'

He narrowed his eyes and said: 'Since when did you start worrying about mankind? Did you ever care for my world? I wanted you to leave it alone, just as it was, free of every damn pollution. But you destroyed it with smog, radiation, radioactive ash, and your experiments. Where is my school now, my village, my two sons? Where are they? You have accounted for them all. Your greed knew no bounds. You always wanted to hoard up much more than what you actually needed. You did not even spare the seas and the mountains. For you these were merely places to be used for installing nuclear devices to spy on the movements of your enemies. Why should I allow myself or my wife to be exploited for the sake of mankind's survival? And come to think of it, when did you ever care for man's survival?'

The men outside were constantly in touch, by radio, with their respective governments, seeking advice and also briefing them on the latest situation, moment by moment. They came to know that the schoolmaster's wife was ill, a piece of news which made them jump with joy because they had the medicine on them which could cure her. But the schoolmaster turned down their offer of help.

The day drew to a close.

The visitors began to prepare for settling down on the hilltop for the night. They kept getting strange messages every now and then. For instance, the astronauts who were about to leave for some planet asked before lifting off: 'Has the woman agreed?'

A cricket commentary which was being relayed through a communications satellite and followed closely by a number of listeners stopped abruptly. Someone brought his mouth close to the mike and said to the commentator: 'Shut up! Let me talk! Has the woman agreed?' And on being answered in the negative he uttered a thanks and returned the mike to the commentator.

With nationals from various countries converging there, the hilltop became fairly populated in a matter of days. Their sole task was to keep the rest of the world informed about the woman's illness and the decision of her husband. The condition of the woman was worsening rapidly and neither she nor her husband were willing to quit their cave.

Outside a heated debate often erupted among the representatives of the various countries as they bickered to apportion the responsibility for the catastrophe and to determine which country was the first to carry out experiments in genetics.

The representative of a small country said proudly: 'Thank God the initiative was not taken by us. We, who belong to the eastern nations, still cherish our spiritual values.'

The representative of a major power, fuming with anger, looked at him and said: 'Whose fancy do you wish to tickle? What about the country in your neighbourhood you were always hell-bent on crushing? A country as eastern as you are and possessing as many spiritual values as you do.'

A delegate belonging to a black country piped up: 'We never wanted all this to happen. We could have made short work of our enemies any time with bows, arrows, and spears.'

'You never wanted all this to happen? You belong to one of those petty troublemakers who kept pestering us: Look alive! Any more neglect and the other superpower will overrun the whole world. Which door did you not knock at to beg for arms?

You and your spiritual values and your easternness! Were these
values you stood for or bogus beliefs?'

While these men were busy squabbling with each other, the
schoolmaster came out and yanked at the curtain made of skin
to remove it from the cave's mouth. Inside the woman could be
seen lying on a heap of skins. The schoolmaster went back into
the cave and when he reemerged he had a spade in his hand. At
a little distance from the cave he began to dig.

Someone asked from the Space Research Center via the
communications satellite: 'Any news?'

The radio operator said: 'I think the woman died last night.'

There was silence for a while at the other end. Then the
person who had sought the information said inanely: 'That
means she did not agree to take the medicine?'

The radio operator said snappily: 'Of course.'

* * *

The hole the schoolmaster was digging took on a distinct shape
by now. It was six feet long. Inside the cave the woman lay
completely still. Once again someone called from the Space
Research Center via the communications satellite:

'Are you sure the woman is dead?'

'Absolutely,' the operator said.

The person who had called asked: 'What are you going to do
with the new medicine?'

'Don't know. We haven't given it a thought so far,' the
representative of the Save Mankind Organization said.

The man on the other side of the line said: 'Pass it on to us.'

'What for?' the representative of the Save Mankind
Organization asked.

'We know what should be done with it,' the caller from the
Space Research Center said.

Only silence greeted his remark.

After a while the man from the Space Research Center added: 'We will place it in the capsule fitted to the nose of the rocket. It will go off into space along with the rocket.'

* * *

With the help of his spade the schoolmaster kept flinging out the earth from the grave he was digging. He worked slowly, pausing every now and then. He took no notice of the noise the helicopters made as they flew past overhead and went on talking to himself: 'Dear earth, how good you still are and how beautiful. You are so beautiful that I am ready today to yield to you my most beloved possession, one which I denied to your enemies.'

—*Translated by Muhammad Salim-ur-Rahman*